AF335680

# Vibration of Nearly Periodic Structures and Mistuned Bladed Rotors

This is the first comprehensive volume on nearly periodic structures and mistuned blade vibration. Alok Sinha presents fundamental concepts and state-of-the-art techniques in the analysis of free and forced response of a nearly periodic structure, weaving together his own work (covering thirty-five years of research in this field) with works by other researchers. He also discusses similarities between tools used in bladed rotor analysis and condensed matter physics.

Specific subjects covered include:

- Reasons behind mode localization
- Reasons behind amplitude amplification of steady-state response
- State-of-the-art computational techniques for mistuned bladed rotors including multistage rotors
- Identification of mistuning from measured response
- Vibration localization in linear atomic chains
- Analysis of two-dimensional periodic structures

**Alok Sinha** is Professor of Mechanical Engineering at Pennsylvania State University. He also has served as visiting professor of aeronautics and astronautics at Massachusetts Institute of Technology and Stanford University. His areas of teaching and research include vibration, control systems, jet engines, robotics, neural networks, and nanotechnology. He is a Fellow of the American Society of Mechanical Engineers and the American Association for the Advancement of Science, and an Associate Fellow of AIAA. He has received a NASA Certificate of Recognition for Significant Contribution to the Space Shuttle Microgravity Mission. He is the author of *Linear Systems: Optimal and Robust Control* (2007) and *Vibration of Mechanical Systems* (2010). He has also served as an associate editor of the *ASME Journal of Dynamic Systems, Measurement and Control*, the *ASME Journal of Turbomachinery*, and the *AIAA Journal*.

# Vibration of Nearly Periodic Structures and Mistuned Bladed Rotors

ALOK SINHA

Pennsylvania State University

**CAMBRIDGE**
UNIVERSITY PRESS

University Printing House, Cambridge CB2 8BS, United Kingdom

One Liberty Plaza, 20th Floor, New York, NY 10006, USA

477 Williamstown Road, Port Melbourne, VIC 3207, Australia

4843/24, 2nd Floor, Ansari Road, Daryaganj, Delhi – 110002, India

79 Anson Road, #06-04/06, Singapore 079906

Cambridge University Press is part of the University of Cambridge.

It furthers the University's mission by disseminating knowledge in the pursuit of
education, learning, and research at the highest international levels of excellence.

www.cambridge.org
Information on this title: www.cambridge.org/9781107188990
DOI: 10.1017/9781316986806

© Alok Sinha 2017

This publication is in copyright. Subject to statutory exception
and to the provisions of relevant collective licensing agreements,
no reproduction of any part may take place without the written
permission of Cambridge University Press.

First published 2017

Printed in the United States of America by Sheridan Books, Inc.

*A catalogue record for this publication is available from the British Library.*

*Library of Congress Cataloging-in-Publication Data*
Names: Sinha, Alok, 1956–
Title: Vibration of nearly periodic structures and mistuned bladed rotors / Alok Sinha,
Pennsylvania State University.
Description: Cambridge, United Kingdom; New York, NY, USA: Cambridge University Press, 2017.
Identifiers: LCCN 2016049748 | ISBN 9781107188990 (hardback)
Subjects: LCSH: Rotors – Vibration – Mathematical models. | Turbines – Blades.
Classification: LCC TJ1058.S554 2017 | DDC 621.8/2–dc23
LC record available at https://lccn.loc.gov/2016049748

ISBN 978-1-107-18899-0 Hardback

Cambridge University Press has no responsibility for the persistence or accuracy of URLs
for external or third-party Internet Web sites referred to in this publication and does not
guarantee that any content on such Web sites is, or will remain, accurate or appropriate.

To my
Granddaughter: Meera
Daughter: Swarna
Daughter and Son-in-Law: Divya and Vishal
Wife: Hansa

# Contents

# Acronyms

| | |
|---|---|
| CMM: | Coordinate Measurement Machine |
| CMS: | Component Mode Synthesis |
| FE: | Finite Element |
| FEM: | Finite Element Model |
| FMM: | Fundamental Model of Mistuning |
| FOD: | Foreign Object Damage |
| IBR: | Integrally Bladed Rotor |
| KL: | Karhunen-Loeve |
| LFT: | Linear Fractional Transformation |
| LMI: | Linear Matrix Inequality |
| MAC: | Modal Assurance Criterion |
| MDA: | Modal Domain Analysis |
| MMDA: | Modified Modal Domain Analysis |
| NOC: | Number of Occurrences |
| POD: | Proper Orthogonal Decomposition |
| SFMM: | Single Family Mode Model |
| SISO: | Single Input/Single Output |
| SNM: | Subset of Nominal Modes |
| SVD: | Singular Value Decomposition |

# Preface

Bladed disk is an important part of steam and gas turbines. Each blade is designed to be identical. Therefore, the structure of a bladed disk can be described as a periodic structure where a bladed disk can be divided into identical sectors. In practice, these structures are nearly periodic because of manufacturing tolerances. This loss of cyclic symmetry can lead to dramatic change in the dynamic behavior. For free vibration, mode shapes can change a great deal and there can be mode localization phenomenon. For forced response, the amplitude of one blade can be three times higher than the value based on exact periodic structure. There are thousands of papers written on this topic in the last fifty years, some of the earliest papers being Whitehead (1966), Wagner (1967), and Ewins (1969). Aspects include reasons for vibration amplification, reduced-order modeling, intentional mistuning, probabilistic analysis, robust control perspective, efficient computational procedure, and nature of excitation. This monograph presents fundamental concepts and issues in the analysis of mistuned vibration. The author has been doing research in this area for more than thirty-five years. Here, he has weaved together works done by him and other researchers.

Periodic structures are ubiquitous not only in engineering sciences but also in physical and life sciences. As mentioned before, they refer to repeating patterns and wave localization because of infinitesimally small amounts of nonlinearities or disorder that are inevitable. This Nobel Prize–winning phenomenon is known as Anderson localization (Anderson, 1958) in condensed matter physics, where it refers to electron localization. Carbon nanotubes and graphenes can have localization of both electrical and mechanical waves. In life sciences, some parts of the DNA double helix open locally at high temperatures and form the so-called denaturation bubbles, which play a role in biological function. Localized transient waves are found in cortical spreading depression in migraine. It is hoped that concepts explained in this book will help researchers in nanotechnology and life sciences as well.

In Chapter 1, free vibration characteristics of nearly periodic structures are explained on the basis of simple models that have been extensively used in the literature. First, existence of repeated eigenvalues and nonuniqueness of eigenvectors is shown for a perfectly tuned system. Then, a computational algorithm is presented to obtain mistuned frequencies and mode shapes by defining the concept of differential eigenvectors. The case of small coupling stiffness is considered to explain the mode localization phenomenon. Next, the natural frequency veering phenomenon, which is widely used in the literature, is explained. This is followed by the analysis demonstrating stabilizing effects of mistuning on aeroelastic instability or flutter of a rotor stage. Lastly, analysis and results on vibration localization in atomic chains are presented.

In Chapter 2, forced vibration characteristics of nearly periodic structures are explored again on the basis of simple models that have been extensively used in the literature. The nature of excitation is considered to be one of the Fourier components of the periodic force caused by circumferential pressure field and constant rotational speed of the bladed rotor. These excitations are characterized by identical amplitude for each blade and constant interblade phase angles between adjacent blades corresponding to one of the natural modes of vibration of the perfectly tuned system. First, it is shown how only a single mode of a perfectly tuned system gets excited. An interesting connection with pole/zero cancellations that is related to loss of observability and controllability in control theory is established. Then, in the presence of mistuning, the phenomenon of energy transfer to many modes is explained as the reason of amplitude amplification. The statistical distribution of the peak maximum amplitude, maximum over all blades and all excitation frequencies, are examined using Monte Carlo simulation. Then, a classical analytical result on the maximum amplitude amplification along with recent numerical algorithms based on infinity norm and linear robust control theory are presented. Last, analytical approaches to compute the statistical distribution of amplitudes are presented on the basis of known statistical distribution of mistuning parameters.

In Chapter 3, reduced-order models of a bladed rotor are presented. These reduced-order models are needed for Monte Carlo simulations to determine the response statistics as the full model is extremely large. First, theory behind the cyclic symmetry analysis, a standard tool in commercially available Finite Element code, is presented, which renders the analysis of a full tuned rotor by an equivalent sector analysis resulting in a computational saving. Similar technique, known as Bloch's theorem, is presented for two-dimensional periodic structures such as graphene and carbon nanotubes. It should be noted that the cyclic symmetry is lost in the presence of mistuning, and equivalent sector analysis is not valid. First, subset of nominal modes (SNMs), single family mode model (SFMM), and component mode synthesis (CMS) approaches are presented to develop reduced-order models. Mistuning identification algorithm associated with SFMM approach is also presented. These techniques are based on the representation of mistuning as variations in blades' natural frequencies alone, known as frequency mistuning. They are unable to yield accurate results in the case of geometric mistuning, which refers to blade-to-blade variations in geometry resulting in simultaneous and dependent perturbations in both mass and stiffness matrices. The breakthrough method modified modal domain analysis (MMDA), which has been developed by the author of this monograph, is presented in detail. Numerical examples are presented to highlight the accuracies of MMDA. Next, algorithm to estimate geometric mistuning from measured data is presented. Lastly, MMDA-like algorithm is developed for the reduced-order model of multistage rotors.

I want to thank my doctoral students who have worked with me on mistuning related topics: S. Chen, D. Cha, Y. Bhartiya, and V. Vishwakarma. I thank Dr. Om Sharma, United Technologies, for his encouragement. I also thank Pratt and Whitney, Air Force Research Laboratory, and GUIde consortium at Duke University for supporting my MMDA research.

# 1 Fundamentals of Free Vibration of a Rotationally Periodic Structure

First, repeated nature of eigenvalues for a perfectly tuned system and the phenomenon of mode-splitting due to mistuning are presented. Next, mode localization and frequency veering phenomenon in a mistuned bladed disk vibration are shown. Then, beneficial effects of mistuning on the aeroelastic instability or flutter are presented. Lastly, vibration localization phenomenon in an atomic chain with disorder (mistuning) is presented.

## 1.1 Basic Mistuned Model for Free Vibration

The model shown in Figure 1.1.1 considers only one mode of vibration per blade. Modal mass and stiffness of each blade are represented by $m_i$ and $k_i$, respectively. These quantities are expressed as

$$m_i = m_t + \delta m_i; \ i = 1, 2, ..., n \tag{1.1.1}$$

$$k_i = k_t + \delta k_i; \ i = 1, 2, ..., n \tag{1.1.2}$$

For a mistuned system, $\delta m_i \neq 0$ and $\delta k_i \neq 0$. The structural coupling between adjacent blades due to the disk flexibility is represented by a spring with stiffness $k_c$. The governing system of differential equations is represented by

$$m_i \ddot{x}_i + k_i x_i + k_c(x_i - x_{i+1}) + k_c(x_i - x_{i-1}) = 0; \ i = 1, 2, ..., n \tag{1.1.3}$$

For $n$ masses in the rotationally periodic structure, $i + 1 = 1$ when $i = n$, and $i - 1 = n$ when $i = 1$.

In matrix form, Equation (1.1.3) can be written as

$$M\ddot{\mathbf{x}} + K\mathbf{x}(t) = \mathbf{0} \tag{1.1.4}$$

where

$$M = M_t + \Delta M \tag{1.1.5}$$

$$K = K_t + \Delta K \tag{1.1.6}$$

$$M_t = m_t I_n \tag{1.1.7}$$

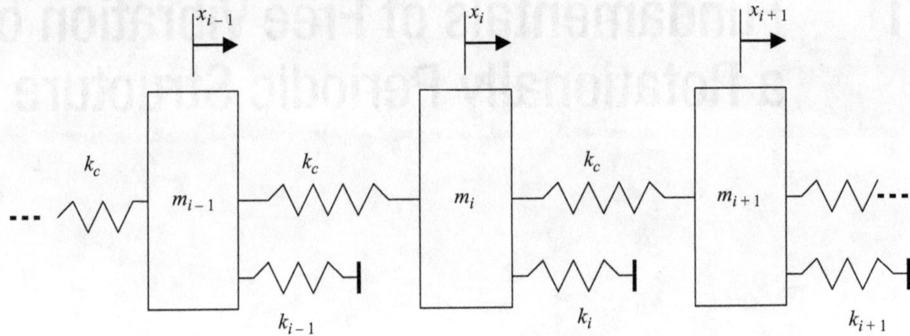


**Figure 1.1.1.**   Undamped mistuned rotationally periodic structure.

$$K_t = \begin{bmatrix} k_t + 2k_c & -k_c & 0 & . & . & -k_c \\ -k_c & k_t + 2k_c & -k_c & 0 & . & 0 \\ 0 & -k_c & k_t + 2k_c & -k_c & - & 0 \\ . & . & . & . & . & . \\ -k_c & 0 & . & 0 & -k_c & k_t + 2k_c \end{bmatrix} \tag{1.1.8}$$

$$\Delta K = \begin{bmatrix} \delta k_1 & & & & \\ & \delta k_2 & & & \\ & & \ddots & & \\ & & & \delta k_{n-1} & \\ & & & & \delta k_n \end{bmatrix} \tag{1.1.9}$$

$$\Delta M = \begin{bmatrix} \delta m_1 & & & & \\ & \delta m_2 & & & \\ & & \ddots & & \\ & & & \delta m_{n-1} & \\ & & & & \delta m_n \end{bmatrix} \tag{1.1.10}$$

## 1.2        Undamped Free Vibration: Tuned System

For a perfectly tuned system,

$$\delta m_i = 0; \;\; \delta k_i = 0; \;\; i = 1, 2, \ldots, n \tag{1.2.1}$$

Equation (1.1.4) reduces to

$$M_t \ddot{\mathbf{x}} + K_t \mathbf{x}(t) = 0 \tag{1.2.2}$$

### 1.2.1     Eigenvalues and Eigenvectors

The natural frequencies ($\omega$) and the modal vectors (**u**) are obtained by the solution of the following eigenvalue/eigenvector problem (Appendix A):

$$K_t \mathbf{u} = \omega^2 m_t \mathbf{u} \tag{1.2.3}$$

Because the matrix $(K_t - \omega^2 m_t I_n)$ is circulant, results in Appendix B yield

$$\det(K_t - \omega^2 m_t I_n) = \begin{cases} (k_t - \omega^2 m_t)\displaystyle\prod_{\ell=1}^{\frac{n-1}{2}}(k_t - \omega^2 m_t + 2k_c(1-\cos(\ell\varphi))^2; \quad \text{for odd } n \\[2em] \left[(k_t - \omega^2 m_t)\displaystyle\prod_{\ell=1}^{\frac{n-2}{2}}(k_t - \omega^2 m_t + 2k_c(1-\cos(\ell\varphi))^2\right](k_t - \omega^2 m_t + 4k_c); \quad \text{for even } n \end{cases} \tag{1.2.4}$$

There are $(n-1)/2$ and $(n-2)/2$ repeated natural frequencies for odd and even $n$, respectively. Eigenvectors are given by Equations (B.3) and (B.4), that is,

$$\mathbf{p}_\ell = \frac{1}{\sqrt{nm_t}}[1 \quad e^{j\phi\ell} \quad e^{j2\phi\ell} \quad \cdots \quad e^{j(n-1)\phi\ell}]^T; \; \ell = 0,1,2,...,(n-1); \; j = \sqrt{-1} \tag{1.2.5}$$

where

$$\phi = \frac{2\pi}{n} \tag{1.2.6}$$

Eigenvectors are scaled because (Appendix A)

$$\Phi^H m_t \Phi = I_n \tag{1.2.7}$$

$$\Phi^H K_t \Phi = \Omega^2 \tag{1.2.8}$$

$$\Omega^2 = \begin{bmatrix} \omega_{f1}^2 & & & & \\ & \omega_{f2}^2 & & & \\ & & \ddots & & \\ & & & \omega_{fn-1}^2 & \\ & & & & \omega_{fn}^2 \end{bmatrix} \tag{1.2.9}$$

$$\Phi = \begin{bmatrix} \mathbf{p}_1 & \mathbf{p}_2 & \cdots & \mathbf{p}_{n-1} & \mathbf{p}_n \end{bmatrix} \tag{1.2.10}$$

and $\Phi^H$ is the complex conjugate transpose of the matrix $\Phi$. For a perfectly tuned system, the number of repeated eigenvalues equals $(n-1)/2$ and $(n-2)/2$ for odd and even number of masses $n$. This also implies that the number of unrepeated eigenvalues equals 1 and 2 for odd and even number of blades $n$, respectively. For odd $n$, the eigenvector corresponding to the unrepeated eigenvalue represents 0 degree intermass phase angle tuned mode. For even $n$, the eigenvectors corresponding to unrepeated eigenvalues represent 0 and 180 degrees intermass phase angle tuned modes.

For repeated eigenvalues, eigenvectors are not unique. If $\mathbf{p}_\ell$ and $\mathbf{p}_{n-\ell}$ are two independent eigenvectors corresponding to a repeated eigenvalue $\lambda_\ell$, it can be easily seen that

$$K_t(\alpha\mathbf{p}_\ell + \beta\mathbf{p}_{n-\ell}) = \lambda_\ell m_t(\alpha\mathbf{p}_\ell + \beta\mathbf{p}_{n-\ell}) \tag{1.2.11}$$

In other words, any linear combination of $\mathbf{p}_\ell$ and $\mathbf{p}_{n-\ell}$ is also an eigenvector. Here,

$$\alpha\mathbf{p}_\ell + \beta\mathbf{p}_{n-\ell} = \frac{\alpha}{\sqrt{nm_t}}\begin{bmatrix} 1 \\ e^{j\ell\phi} \\ e^{j2\ell\phi} \\ \vdots \\ e^{j(n-1)\ell\phi} \end{bmatrix} + \frac{\beta}{\sqrt{nm_t}}\begin{bmatrix} 1 \\ e^{-j\ell\phi} \\ e^{-j2\ell\phi} \\ \vdots \\ e^{-j(n-1)\ell\phi} \end{bmatrix} \tag{1.2.12}$$

Equation (1.2.12) can also be written as

$$\alpha\mathbf{p}_\ell + \beta\mathbf{p}_{n-\ell} = \frac{(\alpha+\beta)}{\sqrt{nm_t}}\begin{bmatrix} 1 \\ \cos(\ell\phi) \\ \cos(2\ell\phi) \\ \vdots \\ \cos((n-1)\ell\phi) \end{bmatrix} + j\frac{(\alpha-\beta)}{\sqrt{nm_t}}\begin{bmatrix} 0 \\ \sin(\ell\phi) \\ \sin(2\ell\phi) \\ \vdots \\ \sin((n-1)\ell\phi) \end{bmatrix} \tag{1.2.13}$$

Equation (1.2.13) suggests that two independent eigenvectors for a repeated eigenvalue can also be described as

$$\begin{bmatrix} 1 \\ \cos(\ell\phi) \\ \cos(2\ell\phi) \\ \vdots \\ \cos((n-1)\ell\phi) \end{bmatrix} \quad and \quad \begin{bmatrix} 0 \\ \sin(\ell\phi) \\ \sin(2\ell\phi) \\ \vdots \\ \sin((n-1)\ell\phi) \end{bmatrix}; \ \ell = \begin{cases} 1,2,\cdots,\dfrac{n-1}{2}; & \textit{for odd } n \\[2mm] 1,2,\cdots,\dfrac{n}{2}-1; & \textit{for even } n \end{cases} \tag{1.2.14}$$

An important property to note here is that these eigenvectors are also orthogonal from Equation (B.17), that is,

$$\begin{bmatrix} 1 & \cos(\ell\phi) & \cos(2\ell\phi) & \cdots & \cos((n-1)\ell\phi) \end{bmatrix}\begin{bmatrix} 0 \\ \sin(\ell\phi) \\ \sin(2\ell\phi) \\ \vdots \\ \sin((n-1)\ell\phi) \end{bmatrix} = 0 \tag{1.2.15}$$

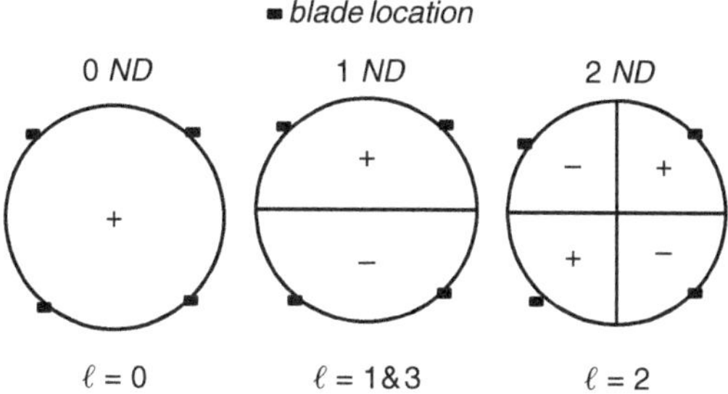


**Figure 1.2.1.** Nodal diameters for a disk with four blades.

The number $\ell$ in Equation (1.2.14) also stands for the number of times the displacement of a mass changes signs as one round of the periodic structure is traversed (see Figure 1.2.1). This number is described as harmonic index. Because of similarities with nodal diameters of a circular disk vibration (Prescott, 1946) where the mode shapes are described by $\cos \ell\theta$ and $\sin \ell\theta$ ($0 \le \theta \le 2\pi$), $\ell$ is also described as the number of nodal diameters. Zero nodal diameter is represented by $\ell = 0$. The maximum numbers of nodal diameters for odd and even $n$ are $(n-1)/2$ and $n/2$, respectively.

## 1.2.2 Traveling Wave Representation

Each modal vector, Equation (1.2.5), can be viewed as a traveling wave. Using Equation (1.2.5),

$$\mathbf{x}(t) = \frac{\alpha}{\sqrt{nm_t}} \begin{bmatrix} 1 \\ e^{j\ell\phi} \\ e^{j2\ell\phi} \\ \vdots \\ e^{j(n-1)\ell\phi} \end{bmatrix} e^{j\omega t} = \frac{\alpha}{\sqrt{nm_t}} \begin{bmatrix} e^{j\omega t} \\ e^{j(\omega t+\ell\phi)} \\ e^{j(\omega t+2\ell\phi)} \\ \vdots \\ e^{j(\omega t+(n-1)\ell\phi)} \end{bmatrix} \tag{1.2.16}$$

Equation (1.2.16) indicates that the phase of sinusoidal vibration of each blade changes by a constant value from blade to blade as the periodic structure is traversed clockwise. This is as if a sinusoidal wave is traveling forward (clockwise).

## 1.2.3 Standing Wave Representation

Forward and backward traveling waves combine to form a standing wave. Using Equation (1.2.12),

$$\mathbf{x}(t) = \frac{\alpha}{\sqrt{nm_t}} \begin{bmatrix} 1 \\ e^{j\ell\phi} \\ e^{j2\ell\phi} \\ \vdots \\ e^{j(n-1)\ell\phi} \end{bmatrix} e^{j\omega t} + \frac{\beta}{\sqrt{nm_t}} \begin{bmatrix} 1 \\ e^{-j\ell\phi} \\ e^{-j2\ell\phi} \\ \vdots \\ e^{-j(n-1)\ell\phi} \end{bmatrix} e^{j\omega t} \tag{1.2.17}$$

Let the coefficients $\alpha$ and $\beta$ be complex conjugates, that is,

$$\alpha = |\alpha| e^{j\psi} \quad \text{and} \quad \beta = \alpha^* = |\alpha| e^{-j\psi} \tag{1.2.18}$$

Then Equation (1.2.17) yields

$$\mathbf{x}(t) = \frac{2|\alpha|}{\sqrt{nm_t}} \begin{bmatrix} \cos\psi \\ \cos(\ell\phi + \psi) \\ \cos(2\ell\phi + \psi) \\ \vdots \\ \cos((n-1)\ell\phi + \psi) \end{bmatrix} (\cos\omega t + j\sin\omega t) \tag{1.2.19}$$

Equivalently, Equation (1.2.19) can be written as

$$\mathbf{x}(t) = \frac{2|\alpha|}{\sqrt{nm_t}} \begin{bmatrix} \cos\psi \\ \cos(\ell\phi + \psi) \\ \cos(2\ell\phi + \psi) \\ \vdots \\ \cos((n-1)\ell\phi + \psi) \end{bmatrix} \cos(\omega t + \pi/4) \tag{1.2.20}$$

Equation (1.2.20) indicates that each blade will vibrate with the same frequency but with different amplitude.

### 1.2.4 Equivalent Single Degree of Freedom Model

An equivalent single degree of freedom of model can be obtained for each harmonic index $\ell$ as the tuned modes (1.2.5) satisfy

$$x_{i+1} = e^{j\phi\ell} x_i \quad \text{and} \quad x_{i-1} = e^{-j\phi\ell} x_i \tag{1.2.21}$$

Substituting Equation (1.2.21) into Equation (1.1.3) for the tuned system,

$$m_t \ddot{x}_i + (k_t + 2k_c)x_i - k_c x_i e^{j\phi\ell} - k_c x_i e^{-j\phi\ell} = 0 \tag{1.2.22}$$

After simplification,

$$m_t \ddot{x}_i + \left( k_t + 4k_c \sin^2\left(\frac{\phi\ell}{2}\right) \right) x_i = 0 \tag{1.2.23}$$

The equivalent single degree of freedom governed by Equation (1.2.23) is shown in Figure 1.2.2.

Tuned natural frequencies are

$$\sqrt{\frac{\left( k_t + 4k_c \sin^2\left(\frac{\phi\ell}{2}\right) \right)}{m_t}}; \ \ell = 0, 1, 2, \ldots, n-1 \tag{1.2.24}$$

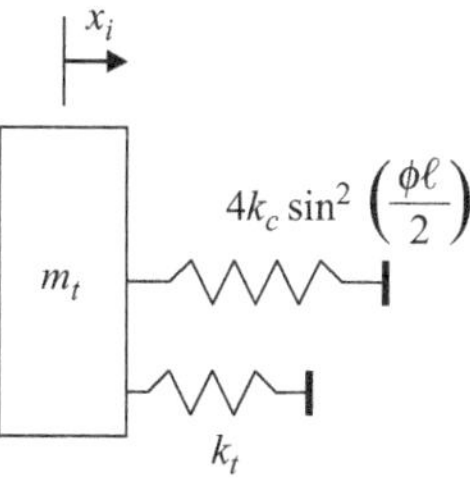


**Figure 1.2.2.** Equivalent single degree of freedom model.

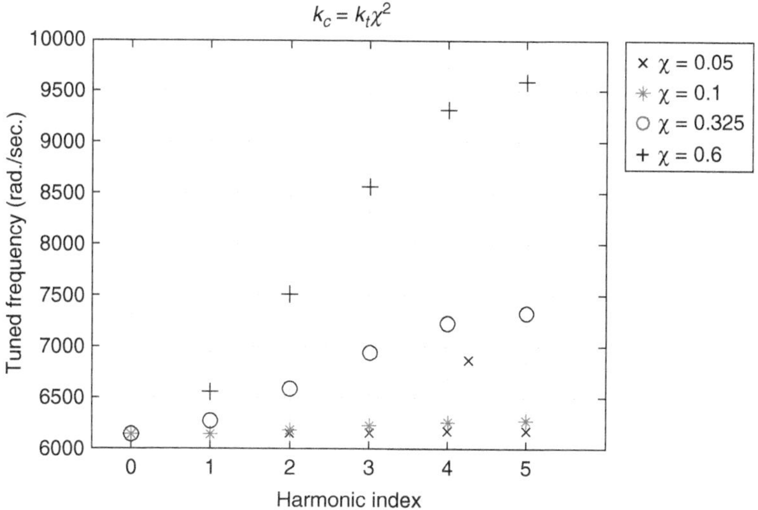


**Figure 1.2.3.** Tuned frequencies ($k_t = 430{,}000\,\text{N/m}$, $m_t = 0.0114\,\text{kg}$).

For a ten mass system ($n = 10$), tuned natural frequencies are shown in Figure 1.2.3 as a function of harmonic indices or nodal diameters for different values of coupling stiffness $k_c$, which is represented by a nondimensional parameter $\chi$ as follows. Define

$$\delta\omega_i^2 = \frac{\delta k_i}{k_t}, \quad \omega_{bt}^2 = \frac{k_t}{m_t}, \quad \omega_c^2 = \frac{k_c}{m_t}; \quad \chi^2 = \frac{\omega_c^2}{\omega_{bt}^2} = \frac{k_c}{k_t} \tag{1.2.25}$$

In this case, Equation (1.1.3) can be written as

$$\ddot{x}_i + \omega_{bt}^2[(1 + 2\chi^2 + \delta\omega_i^2)x_i + \chi^2 x_{i+1} - \chi^2 x_{i-1}] = 0; \; i = 1,2,\ldots,n \tag{1.2.26}$$

When $n = 10$, the harmonic index $\ell$ ranges from 0 to 5, with $\ell = 0$ and $\ell = 5$ representing unrepeated frequencies corresponding to 0 and 180 degrees interblade phase angles, Figure 1.2.3. And, the values of $\ell$ between 1 and 4 represent repeated frequencies. Here, frequencies are identical for the following pairs of $\ell$ values: (1, 9), (2,8), (3,7), and (4,6). That is why frequencies are only plotted for $\ell$ ranging from 0 to 5.

## 1.3     Undamped Free Vibration: Mistuned System

Equation (1.1.4) can be written as

$$(M_t + \Delta M)\ddot{\mathbf{x}} + (K_t + \Delta K)\mathbf{x}(t) = 0 \tag{1.3.1}$$

The corresponding eigenvalue/eigenvector problem (Appendix A) is

$$(K_t + \Delta K)\mathbf{v} = \lambda(M_t + \Delta M)\mathbf{v} \tag{1.3.2}$$

Because of perturbations in mass and stiffness matrices, repeated eigenvalues for the tuned system split and the mistuned system has distinct eigenvalues and unique eigenvectors. An important question is how are these eigenvectors related to independent eigenvectors (Sinha, 2006a). Xiangjun and Shijing (1986) have commented that the eigenvector corresponding to a repeated eigenvalue is a discontinuous function of system parameters. Applying results in Andrew and Tan (1998), both eigenvalues and eigenvectors should be analytic with respect to a parameter on which perturbations of mass and stiffness matrices depend. Zhang and Wang (1995) have developed an analytical approach to compute the derivatives of repeated eigenvalues and corresponding eigenvectors of a nondefective matrix. One of their important contributions is to show that there exists a particular linear combination of independent eigenvectors $\mathbf{v}_i$ and $\mathbf{v}_{i+1}$ corresponding to a repeated eigenvalue that is differentiable. However, with respect to an arbitrary choice of this linear combination, the eigenvector corresponding to a repeated eigenvalue is discontinuous as described by Xiangjun and Shijing (1986). Shapiro (1998, 1999) has used a multidimensional Taylor series to compute mistuned eigenvalues. He has also shown that the eigenvalue of a mistuned system is a continuous function of mistuned parameters, and it can appear to be discontinuous because of mode switching. However, none of the cited papers has dealt with the computation of mistuned eigenvectors using Taylor series expansion. Here, the analyses developed by Sinha (2006a) are presented.

### 1.3.1     Derivatives of Eigenvalues and Eigenvectors

#### Repeated Eigenvalues

Let the changes in mass and stiffness matrices be a function of a single parameter $r$. Also, it is assumed that derivatives of mass and stiffness matrices with respect to the parameter $r$ are known. Let $\mathbf{v}_i$ and $\mathbf{v}_{i+1}$ be two independent eigenvectors of a perfectly tuned system corresponding to a repeated eigenvalue $\lambda_0$. Define

$$X = \begin{bmatrix} \mathbf{v}_i & \mathbf{v}_{i+1} \end{bmatrix} \tag{1.3.3}$$

Therefore, linear combinations of these two eigenvectors are described as

$$Z = X\Gamma \tag{1.3.4}$$

where $\Gamma$ is a square matrix of dimension 2 with the following property:

$$\Gamma^{-1} = \Gamma^{T} \tag{1.3.5}$$

Now, from Equations (1.3.2), (1.1.5), and (1.1.6),

$$KZ = MZ\Lambda \tag{1.3.6}$$

where

$$\Lambda = \lambda_0 I_2 \tag{1.3.7}$$

Differentiating Equation (1.3.6) with respect to an independent parameter $r$ and evaluating derivatives at the nominal tuned system,

$$(K - \lambda_0 M)\frac{dZ}{dr} + \left(\frac{dK}{dr} - \lambda_0\frac{dM}{dr}\right)Z = MZ\frac{d\Lambda}{dr} \tag{1.3.8}$$

where

$$\frac{d\Lambda}{dr} = \begin{bmatrix} \dfrac{d\lambda_{0,1}}{dr} & 0 \\ 0 & \dfrac{d\lambda_{0,2}}{dr} \end{bmatrix} \tag{1.3.9}$$

Premultiplying Equation (1.3.8) by $Z^T$,

$$\frac{d\Lambda}{dr} = Z^T\left(\frac{dK}{dr} - \lambda_0\frac{dM}{dr}\right)Z \tag{1.3.10}$$

Substituting Equation (1.3.6) into Equation (1.3.10), and using Equation (1.3.4),

$$X^T\left(\frac{dK}{dr} - \lambda_0\frac{dM}{dr}\right)X\Gamma = \Gamma\frac{d\Lambda}{dr} \tag{1.3.11}$$

Hence, the matrices $\Gamma$ and $\dfrac{d\Lambda}{dr}$ are obtained by solving the eigenvalue/eigenvector problem (1.3.11). Matrix $\Gamma$ represents the combination of eigenvectors that gets uniquely defined for a variation in a parameter. With respect to this combination of eigenvectors, eigenvectors of the mistuned system are a continuous function of the parameter $r$, and their derivatives can be computed as follows.

Substituting Equation (1.3.10) into Equation (1.3.8),

$$(K - \lambda_0 M)\frac{dZ}{dr} = (MZZ^T - I)\left(\frac{dK}{dr} - \lambda_0 \frac{dM}{dr}\right)Z \tag{1.3.12}$$

The dimension of the null space of $(K - \lambda_0 M)$ is 2 for the tuned system, and independent vectors in the null space (Strang, 1988) are columns of the matrix $Z$. Therefore, a general solution of Equation (1.3.12) can be written as

$$\frac{dZ}{dr} = W + ZS \tag{1.3.13}$$

where $W$ is a particular solution of Equation (1.3.12) and $ZS$ is the homogeneous solution where the coefficient matrix $S$ is determined from the second derivatives of eigenvalues. Differentiating Equation (1.3.8) with respect to $r$,

$$\begin{aligned}
(K - \lambda_0 M)\frac{d^2 Z}{dr^2} &+ \left(\frac{d^2 K}{dr^2} - \lambda_0 \frac{d^2 M}{dr^2}\right)Z + 2\left(\frac{dK}{dr} - \lambda_0 \frac{dM}{dr}\right)\frac{dZ}{dr} \\
&= MZ\frac{d^2 \Lambda}{dr^2} + 2M\frac{dZ}{dr}\frac{d\Lambda}{dr} + 2\frac{dM}{dr}Z\frac{d\Lambda}{dr}
\end{aligned} \tag{1.3.14}$$

Premultiplying Equation (1.3.14) by $Z^T$,

$$\begin{aligned}
\frac{d^2 \Lambda}{dr^2} = Z^T\left(\frac{d^2 K}{dr^2} - \lambda_0 \frac{d^2 M}{dr^2}\right)Z &+ 2Z^T\left(\frac{dK}{dr} - \lambda_0 \frac{dM}{dr}\right)\frac{dZ}{dr} \\
&- 2Z^T M\frac{dZ}{dr}\frac{d\Lambda}{dr} - 2Z^T\frac{dM}{dr}Z\frac{d\Lambda}{dr}
\end{aligned} \tag{1.3.15}$$

Substituting Equation (1.3.13) into Equation (1.3.15) and using Equation (1.3.10),

$$2S\frac{d\Lambda}{dr} - 2\frac{d\Lambda}{dr}S + \frac{d^2 \Lambda}{dr^2} = U \tag{1.3.16}$$

where

$$\begin{aligned}
U = Z^T\left(\frac{d^2 K}{dr^2} - \lambda_0 \frac{d^2 M}{dr^2}\right)Z &+ 2Z^T\left(\frac{dK}{dr} - \lambda_0 \frac{dM}{dr}\right)W \\
&- 2Z^T MW\frac{d\Lambda}{dr} - 2Z^T\frac{dM}{dr}Z\frac{d\Lambda}{dr}
\end{aligned} \tag{1.3.17}$$

Let $s_{ij}$ and $u_{ij}$ be elements of matrices $S$ and $U$ in ith row and jth column, respectively. Then, equating off-diagonal elements on both sides of Equation (1.3.16),

$$s_{12} = \frac{u_{12}}{2\left(\dfrac{d\lambda_{0,2}}{dr} - \dfrac{d\lambda_{0,1}}{dr}\right)} \tag{1.3.18}$$

and

$$s_{21} = \frac{u_{21}}{2\left(\dfrac{d\lambda_{0,1}}{dr} - \dfrac{d\lambda_{0,2}}{dr}\right)} \tag{1.3.19}$$

Diagonal elements of the matrix $S$ are obtained from the following normalization condition:

$$\mathbf{z}_i^T M \mathbf{z}_i = 1 \tag{1.3.20}$$

where $\mathbf{z}_i$ is the ith column of the matrix $Z$. Differentiating Equation (1.3.20) with respect to $r$,

$$2\mathbf{z}_i^T M \frac{d\mathbf{z}_i}{dr} + \mathbf{z}_i^T \frac{dM}{dr} \mathbf{z}_i = 0 \tag{1.3.21}$$

From Equation (1.3.13),

$$\frac{d\mathbf{z}_i}{dr} = \mathbf{w}_i + Z\mathbf{s}_i \tag{1.3.22}$$

where $\mathbf{w}_i$ and $\mathbf{s}_i$ are ith column of the matrices $W$ and $S$, respectively. Substituting Equation (1.3.22) into Equation (1.3.21),

$$s_{ii} = -\mathbf{z}_i^T M \mathbf{w}_i - 0.5 \mathbf{z}_i^T \frac{dM}{dr} \mathbf{z}_i ; \ i = 1 \text{ and } 2 \tag{1.3.23}$$

Having obtained the matrix $S$, the second derivative of eigenvalues can be obtained from Equation (1.3.15).

## Unrepeated Eigenvalues

Differentiating Equation (1.3.2) with respect to an independent parameter $r$ (Nelson, 1976);

$$(K - \lambda_i M)\frac{d\mathbf{v}_i}{dr} = \frac{d\lambda_i}{dr} M \mathbf{v}_i - \frac{dK}{dr} \mathbf{v}_i + \lambda_i \frac{dM}{dr} v_i \tag{1.3.24}$$

Premultiplying Equation (1.3.24) by $\mathbf{v}_j^T$,

$$(\lambda_j - \lambda_i)\mathbf{v}_j^T M \frac{d\mathbf{v}_i}{dr} = \frac{d\lambda_i}{dr} \mathbf{v}_j^T M \mathbf{v}_i - \mathbf{v}_j^T \frac{dK}{dr} \mathbf{v}_i + \lambda_i \mathbf{v}_j^T \frac{dM}{dr} \mathbf{v}_i \tag{1.3.25}$$

With $j = i$ and $\mathbf{v}_i^T M \mathbf{v}_i = 1$,

$$\frac{d\lambda_i}{dr} = \mathbf{v}_i^T \frac{dK}{dr} \mathbf{v}_i - \lambda_i \mathbf{v}_i^T \frac{dM}{dr} \mathbf{v}_i \tag{1.3.26}$$

With $j \neq i$ and $\lambda_j \neq \lambda_i$,

$$\mathbf{v}_j^T M \frac{d\mathbf{v}_i}{dr} = \frac{1}{(\lambda_j - \lambda_i)}\left[ -\mathbf{v}_j^T \frac{dK}{dr}\mathbf{v}_i + \lambda_i \mathbf{v}_j^T \frac{dM}{dr}\mathbf{v}_i \right] = q_j \qquad (1.3.27)$$

Since $\mathbf{v}_i$ corresponds to an unrepeated eigenvalue, Equation (1.3.27) represents $(nd - 1)$ equations, where $nd$ is the number of degrees of freedom. Differentiating $\mathbf{v}_i^T M \mathbf{v}_i = 1$,

$$2\mathbf{v}_i^T M \frac{d\mathbf{v}_i}{dr} = -\mathbf{v}_i^T \frac{dM}{dr}\mathbf{v} = q_i \qquad (1.3.28)$$

Combining Equations (1.3.27) and (1.3.28) in the matrix form,

$$\frac{d\mathbf{v}_i}{dr} = P^{-1}\mathbf{q} \qquad (1.3.29)$$

where

$$P = \begin{bmatrix} \mathbf{v}_1 & \cdots & \mathbf{v}_{i-1} & 2\mathbf{v}_i & \mathbf{v}_{i+1} & \cdots & \mathbf{v}_{nd} \end{bmatrix}^T M \qquad (1.3.30)$$

$$\mathbf{q} = \begin{bmatrix} q_1 & \cdots & q_{i-1} & q_i & q_{i+1} & \cdots & q_{nd} \end{bmatrix}^T \qquad (1.3.31)$$

and elements of the vector $\mathbf{q}$ are defined in Equations (1.3.27) and (1.3.28).

Differentiating Equation (1.3.26),

$$\begin{aligned}
\frac{d^2\lambda_i}{dr^2} = {} & \frac{d\mathbf{v}_i^T}{dr}\frac{dK}{dr}\mathbf{v}_i + \mathbf{v}_i^T \frac{dK}{dr}\frac{d\mathbf{v}_i}{dr} + \mathbf{v}_i^T \frac{d^2 K}{dr^2}\mathbf{v}_i \\
& - \lambda_i \left[ \frac{d\mathbf{v}_i^T}{dr}\frac{dM}{dr}\mathbf{v}_i + \mathbf{v}_i^T \frac{dM}{dr}\frac{d\mathbf{v}_i}{dr} + \mathbf{v}_i^T \frac{d^2 M}{dr^2}\mathbf{v}_i \right] - \frac{d\lambda_i}{dr}\mathbf{v}_i^T \frac{dM}{dr}\mathbf{v}_i
\end{aligned} \qquad (1.3.32)$$

and

$$\begin{aligned}
\frac{\partial^2\lambda_i}{\partial q \partial r} = {} & \frac{\partial\mathbf{v}_i^T}{\partial q}\frac{\partial K}{\partial r}\mathbf{v}_i + \mathbf{v}_i^T \frac{\partial K}{\partial r}\frac{\partial\mathbf{v}_i}{\partial q} + \mathbf{v}_i^T \frac{\partial^2 K}{\partial q \partial r}\mathbf{v}_i \\
& - \lambda_i \left[ \frac{\partial\mathbf{v}_i^T}{\partial q}\frac{\partial M}{\partial r}\mathbf{v}_i + \mathbf{v}_i^T \frac{\partial M}{\partial r}\frac{\partial\mathbf{v}_i}{\partial q} + \mathbf{v}_i^T \frac{\partial^2 M}{\partial q \partial r}\mathbf{v}_i \right] - \frac{\partial\lambda_i}{\partial q}\mathbf{v}_i^T \frac{\partial M}{\partial r}\mathbf{v}_i
\end{aligned} \qquad (1.3.33)$$

Equation (1.3.29) can also be expressed (Fox and Kapoor, 1968) as

$$\frac{d\mathbf{v}_i}{dr} = -\frac{\mathbf{v}_i^T \dfrac{dM}{dr}\mathbf{v}_i}{2}\mathbf{v}_i + \sum_{\substack{\ell=1 \\ \ell \neq i}}^{n} \frac{\mathbf{v}_\ell^T \left( \dfrac{dK}{dr} - \lambda_i \dfrac{dM}{dr} \right)\mathbf{v}_i}{\lambda_i - \lambda_\ell}\mathbf{v}_\ell \qquad (1.3.34)$$

## 1.4    Taylor Series Expansion

### 1.4.1    Multidimensional Taylor Series Expansion of an Eigenvalue

Let $\xi_i$; $i = 1, 2, ...., \ell$, be $\ell$ independent random variables describing $\Delta K$ and $\Delta M$. Each of these random variables will have $n$ different values, $\xi_{i1}, \xi_{i2}, ...., \xi_{in}$, in the mistuned system. Assuming that there is only one random variable $\xi_1$,

$$\lambda = \lambda_t + \sum_{j=1}^{n} \frac{\partial \lambda}{\partial \xi_{1j}} \xi_{1j} + \frac{1}{2} \sum_{j=1}^{n} \frac{\partial^2 \lambda}{\partial \xi_{1j}^2} \xi_{1j}^2 + \sum_{i=1}^{n} \sum_{j=i+1}^{n} \frac{\partial^2 \lambda}{\partial \xi_{1i} \partial \xi_{1j}} \xi_{1i} \xi_{1j} + \cdots\cdots \tag{1.4.1}$$

Because of symmetry,

$$\frac{\partial \lambda}{\partial \xi_{11}} = \cdots\cdots = \frac{\partial \lambda}{\partial \xi_{1n}}$$

$$\frac{\partial^2 \lambda}{\partial^2 \xi_{11}} = \cdots\cdots = \frac{\partial^2 \lambda}{\partial^2 \xi_{1n}}$$

$$\frac{\partial^2 \lambda}{\partial \xi_1 \partial \xi_2} = \frac{\partial^2 \lambda}{\partial \xi_2 \partial \xi_3} = \cdots\cdots = \frac{\partial^2 \lambda}{\partial \xi_{n-1} \partial \xi_{n-2}} \tag{1.4.2}$$

$$\frac{\partial^2 \lambda}{\partial \xi_1 \partial \xi_3} = \cdots\cdots\cdots = \frac{\partial^2 \lambda}{\partial \xi_{n-2} \partial \xi_n}$$

$$\frac{\partial^2 \lambda}{\partial \xi_1 \partial \xi_{n-1}}$$

Using the symmetry property (1.4.2), Equation (1.4.1) can be written as

$$\lambda = \lambda_t + \frac{\partial \lambda}{\partial \xi_{11}} \sum_{j=1}^{n} \xi_{1j} + \frac{1}{2} \frac{\partial^2 \lambda}{\partial \xi_{11}^2} \sum_{j=1}^{n} \xi_{1j}^2 + \frac{\partial^2 \lambda}{\partial \xi_{11} \partial \xi_{12}} \sum_{i=1}^{n-1} \xi_{1i} \xi_{1(i+1)}$$
$$+ \frac{\partial^2 \lambda}{\partial \xi_{11} \partial \xi_{13}} \sum_{i=1}^{n-2} \xi_{1i} \xi_{1(i+2)} + \cdots\cdots + \frac{\partial^2 \lambda}{\partial \xi_{11} \partial \xi_{1(n-1)}} \xi_{11} \xi_{1(n-1)} + \cdots\cdots \tag{1.4.3}$$

There are more symmetry properties in a periodic structure like tuned bladed disk:

$$\frac{\partial^2 \lambda}{\partial \xi_{11} \partial \xi_{12}} = \frac{\partial^2 \lambda}{\partial \xi_{11} \partial \xi_{1n}}$$

$$\frac{\partial^2 \lambda}{\partial \xi_{11} \partial \xi_{13}} = \frac{\partial^2 \lambda}{\partial \xi_{11} \partial \xi_{1(n-1)}} \tag{1.4.4}$$

$$\vdots$$

$$\frac{\partial^2 \lambda}{\partial \xi_{11} \partial \xi_{1(h-1)}} = \frac{\partial^2 \lambda}{\partial \xi_{11} \partial \xi_{1(h+1)}} \quad \text{where} \quad h = n/2 \quad \text{for even} \quad n$$

or

$$\frac{\partial^2 \lambda}{\partial \xi_{11} \partial \xi_{1(h-1)}} = \frac{\partial^2 \lambda}{\partial \xi_{11} \partial \xi_{1(g+1)}} \quad \text{where} \quad h = (n+1)/2 \text{ and } g = (n+3)/2 \text{ for odd } n$$

The first equation in (1.4.4) comes from the fact that $\partial \xi_{11} \partial \xi_{12}$ and $\partial \xi_{1n} \partial \xi_{11}$ are products of changes in two consecutive random variables in a circular chain. Similarly, the second equation results from the fact that changes in first ($\xi_{11}$) and third ($\xi_{13}$) random variables will bring the same effect as the changes in $(n-1)th$ ($\xi_{1(n-1)}$)and first ($\xi_{11}$) random variables in a periodic structure. Other equations in (1.4.4) can be derived by a similar logic. From Equations (1.4.3) and (1.4.4), for odd $n$,

$$\lambda = \lambda_t + \frac{\partial \lambda}{\partial \xi_{11}} \sum_{j=1}^{n} \xi_{1j} + \frac{1}{2} \frac{\partial^2 \lambda}{\partial \xi_{11}^2} \sum_{j=1}^{n} \xi_{1j}^2 + \frac{\partial^2 \lambda}{\partial \xi_{11} \partial \xi_{12}} \sum_{i=1}^{n} \xi_{1i} \xi_{1(i+1)}$$
$$+ \frac{\partial^2 \lambda}{\partial \xi_{11} \partial \xi_{13}} \sum_{i=1}^{n} \xi_{1i} \xi_{1(i+2)} + \cdots\cdots\cdots + \frac{\partial^2 \lambda}{\partial \xi_{11} \partial \xi_{1h}} \sum_{i=1}^{n} \xi_{1i} \xi_{1(i+h-1)} + \cdots \tag{1.4.5}$$

with $h = (n+1)/2$.

And, for even $n$,

$$\lambda = \lambda_t + \frac{\partial \lambda}{\partial \xi_{11}} \sum_{j=1}^{n} \xi_{1j} + \frac{1}{2} \frac{\partial^2 \lambda}{\partial \xi_{11}^2} \sum_{j=1}^{n} \xi_{1j}^2 + \frac{\partial^2 \lambda}{\partial \xi_{11} \partial \xi_{12}} \sum_{i=1}^{n} \xi_{1i} \xi_{1(i+1)} + \frac{\partial^2 \lambda}{\partial \xi_{11} \partial \xi_{13}} \sum_{i=1}^{n} \xi_{1i} \xi_{1(i+2)}$$
$$+ \cdots\cdots\cdots + \frac{\partial^2 \lambda}{\partial \xi_{11} \partial \xi_{1(h-1)}} \sum_{i=1}^{n} \xi_{1i} \xi_{1(i+h-1)} + \frac{\partial^2 \lambda}{\partial \xi_{11} \partial \xi_{1h}} \sum_{i=1}^{q} \xi_{1i} \xi_{1(i+h)} + \cdots\cdots \tag{1.4.6}$$

with $h = n/2$.

For a mistuned system, it is quite typical to have mistuning with zero mean, that is, $\sum_{j=1}^{n} \xi_{1j} = 0$. Therefore, mistuned eigenvalues will depend on second-order terms in the multidimensional Taylor series expansion (1.4.5) or (1.4.6). For an unrepeated eigenvalue, required derivatives can be evaluated by Equations (1.3.32) and (1.3.33). For a repeated eigenvalue, pure second-order derivatives can be obtained from Equation (1.3.15), and there will be a need to develop a similar analytical expression for mixed second-order partial derivatives. It should be noted that second-order derivatives of mass and stiffness matrices are present in Equations (1.3.15), (1.3.32), and (1.3.33). While using a commercially available software like ANSYS or NASTRAN, derivatives of mass and stiffness matrices with respect to mistuned parameters may have to be evaluated numerically by finite differences. In this case, numerical efforts can be quite excessive because of many second-order terms. Therefore, a unidirectional Taylor series expansion of an eigenvalue/eigenvector is developed next.

### 1.4.2  Unidirectional Taylor Series Expansion of an Eigenvalue and a "Discontinuous" Eigenvector

According to the theory presented by Andrew and Tan (1998), and observations from numerical results to be discussed in the next section, each eigenvalue, irrespective of being repeated or unrepeated, and the corresponding eigenvector are continuous along any radial direction in mistuning parameters space. For a single random variable case, let the mistuning parameters be:

$$\xi_{11}, \xi_{12}, \ldots\ldots, \xi_{1n} \tag{1.4.7}$$

Based on values of these parameters for a mistuned bladed disk, the following vector in the parameter space can be defined:

$$\mathbf{p} = r\boldsymbol{\chi} \tag{1.4.8}$$

where

$$\mathbf{p} = \begin{bmatrix} \xi_{11} & \xi_{12} & \cdot & \cdot & \xi_{1n} \end{bmatrix}^T \tag{1.4.9}$$

$$\boldsymbol{\chi} = \begin{bmatrix} \chi_{11} & \chi_{12} & \cdot & \cdot & \chi_{1n} \end{bmatrix}^T \tag{1.4.10}$$

$$r = \left[ \sum_{i=1}^{n} \xi_{1i}^2 \right]^{0.5} \tag{1.4.11}$$

Note that $\boldsymbol{\chi}$ is a unit vector. Then, derivatives of mass and stiffness matrices along the vector $\mathbf{p}$ can be calculated as follows:

$$\frac{dM}{dr} = \sum_{i=1}^{n} \frac{\partial M}{\partial \xi_{1i}} \chi_{1i} \tag{1.4.12}$$

and

$$\frac{dK}{dr} = \sum_{i=1}^{n} \frac{\partial K}{\partial \xi_{1i}} \chi_{1i} \tag{1.4.13}$$

These derivatives of mass and stiffness matrices are required for the computation of the derivatives of eigenvalues and eigenvectors. Defining Taylor series expansions of eigenvalues and eigenvectors in terms of a perturbation along the vector $\mathbf{p}$:

$$\lambda = \lambda_t + \frac{d\lambda}{dr} r + \frac{1}{2} \frac{d^2\lambda}{dr^2} r^2 + \cdots\cdots\cdots \tag{1.4.14}$$

and

$$\mathbf{v} = \mathbf{v}_t + \frac{d\mathbf{v}}{dr} r + \frac{1}{2} \frac{d^2\mathbf{v}}{dr^2} r^2 + \cdots\cdots\cdots \tag{1.4.15}$$

**Table 1.4.1.** Natural frequencies (Sinha, 2006a) of the tuned system (rad./sec.)

| | |
|---|---|
| $\omega_1$ | 6141.6 |
| $\omega_2$ | 6264.3 |
| $\omega_3$ | 6264.3 |
| $\omega_4$ | 6574.7 |
| $\omega_5$ | 6574.7 |
| $\omega_6$ | 6939.2 |
| $\omega_7$ | 6939.2 |
| $\omega_8$ | 7220.6 |
| $\omega_9$ | 7220.6 |
| $\omega_{10}$ | 7325.3 |

Eigenvalues and eigenvectors of the mistuned bladed disk are computed by substituting the value of $r$, Equation (1.4.11), in Equations (1.4.14) and (1.4.15). It should be noted that $\lambda_t$ and $\mathbf{v}_t$ are the eigenvalue and the eigenvector of the tuned system, respectively. In the case of a repeated eigenvalue, $\mathbf{v}_t$ is the differential eigenvector obtained from the solution of Equation (1.3.11).

## 1.4.3    Numerical Results

Deviation in the stiffness matrix is described as

$$\Delta K = diag[\, \xi_{11} \quad \xi_{12} \quad \cdot \quad \cdot \quad \xi_{1n}] \tag{1.4.16}$$

and

$$\xi_{1i} = k_i - k_t; \qquad i = 1, 2, ...., n \tag{1.4.17}$$

The mass $m_t$ and stiffness of the tuned system $k_t$ are 0.0114 kg and 430000 N/m, respectively. With the coupling stiffness $k_c$ = 45430 N/m, natural frequencies and eigenvectors of the tuned system are presented in Tables 1.4.1 and 1.4.2 for $n = nd = 10$. Figures 1.4.1–1.4.3 show variations in the unrepeated natural frequency $\omega_1$ and repeated natural frequencies $\omega_2$ and $\omega_3$ as a function of changes in the stiffnesses of first and second blades, $\xi_{11}$ and $\xi_{12}$. While the frequency $\omega_1$ is continuous along any direction, frequencies $\omega_2$ and $\omega_3$ are continuous along radial directions, but seem to be discontinuous along circumferential directions. This stems from the switching of frequencies across the tuned system. In any radial direction, the higher (lower) $\omega_2$ and $\omega_3$ values correspond to the lower (higher) $\omega_2$ and $\omega_3$ values across the tuned system. This fact has already been included in Figures 1.4.2 and 1.4.3. However, frequencies are also continuous along circumferential directions, which is evident when both Figures 1.4.2 and 1.4.3 are viewed together. Edges A and B in Figure 1.4.2 correspond to edges A and B in Figure 1.4.3. Counterparts of edges

**Table 1.4.2.** Modal vectors (Sinha, 2006a) of the tuned system ($r = 0$)

| | $v_1$ | $v_2$ | $v_3$ | $v_4$ | $v_5$ | $v_6$ | $v_7$ | $v_8$ | $v_9$ | $v_{10}$ |
|---|---|---|---|---|---|---|---|---|---|---|
| $x_1$ | 2.9617 | 2.4620 | 3.3886 | 3.9835 | −1.2943 | 1.2943 | 3.9835 | 3.3886 | 2.4620 | −2.9617 |
| $x_2$ | 2.9617 | 3.9835 | 1.2943 | 2.4620 | 3.3886 | 3.3886 | −2.4620 | −1.2943 | −3.9835 | 2.9617 |
| $x_3$ | 2.9617 | 3.9835 | −1.2943 | −2.4620 | 3.3886 | −3.3886 | −2.4620 | −1.2943 | 3.9835 | −2.9617 |
| $x_4$ | 2.9617 | 2.4620 | −3.3886 | −3.9835 | −1.2943 | −1.2943 | 3.9835 | 3.3886 | −2.4620 | 2.9617 |
| $x_5$ | 2.9617 | −0.0000 | −4.1885 | 0.0000 | −4.1885 | 4.1885 | 0.0000 | −4.1885 | −0.0000 | −2.9617 |
| $x_6$ | 2.9617 | −2.4620 | −3.3886 | 3.9835 | −1.2943 | −1.2943 | −3.9835 | 3.3886 | 2.4620 | 2.9617 |
| $x_7$ | 2.9617 | −3.9835 | −1.2943 | 2.4620 | 3.3886 | −3.3886 | 2.4620 | −1.2943 | −3.9835 | −2.9617 |
| $x_8$ | 2.9617 | −3.9835 | 1.2943 | −2.4620 | 3.3886 | 3.3886 | 2.4620 | −1.2943 | 3.9835 | 2.9617 |
| $x_9$ | 2.9617 | −2.4620 | 3.3886 | −3.9835 | −1.2943 | 1.2943 | −3.9835 | 3.3886 | −2.4620 | −2.9617 |
| $x_{10}$ | 2.9617 | 0 | 4.1885 | 0 | −4.1885 | −4.1885 | 0 | −4.1885 | 0 | 2.9617 |

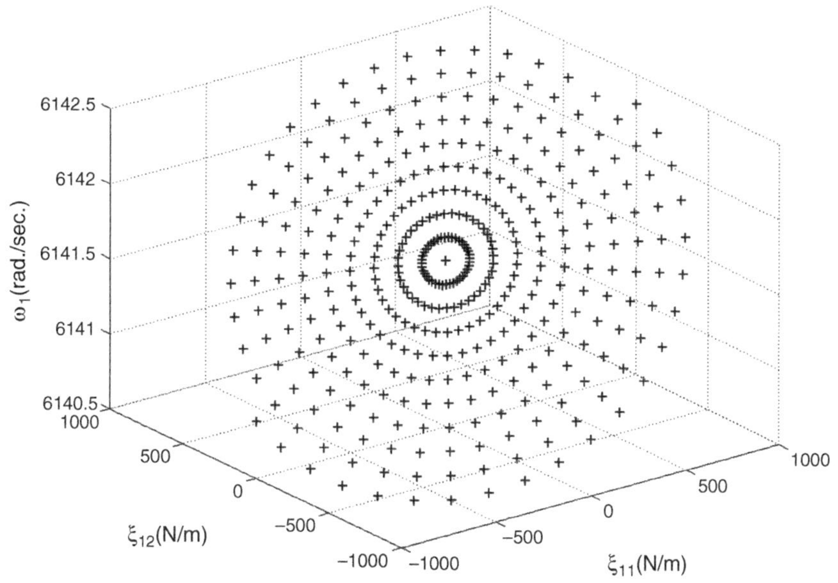


**Figure 1.4.1.** $\omega_1$ as a function of $\xi_{11}$ and $\xi_{12}$ (Sinha, 2006a).

C and D in Figure 1.4.2 are also present in Figure 1.4.3. Therefore, a multidimensional Taylor series can be applied to both unrepeated and repeated natural frequencies as both pure and mixed partial derivatives exist. However, as explained earlier, a unidirectional Taylor series may be preferred because of computational convenience and efficiency.

Using Equation (1.3.11), differentiable eigenvectors have been computed for repeated eigenvalues. For example,

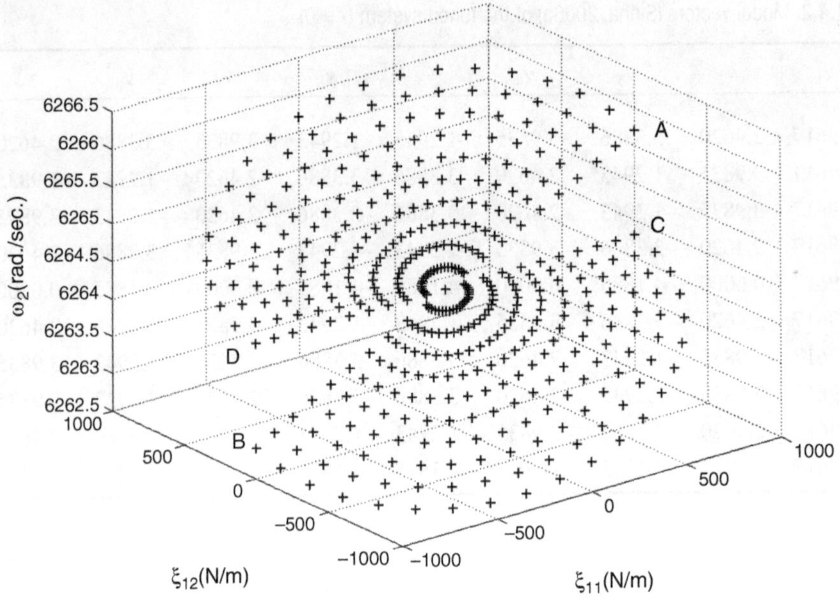

**Figure 1.4.2.** $\omega_2$ as a function of $\xi_{11}$ and $\xi_{12}$ in radial directions (Sinha, 2006a).

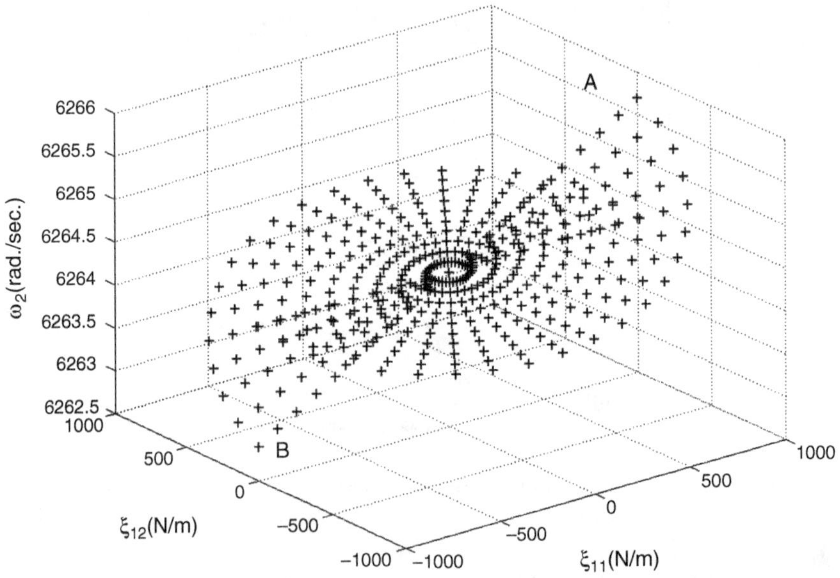

**Figure 1.4.3.** $\omega_3$ as a function of $\xi_{11}$ and $\xi_{12}$ in radial directions (Sinha, 2006a).

differential eigenvectors corresponding to $\omega_4$ and $\omega_5$ are presented in Table 1.4.3 for the unit mistuning vector $\chi$, Table 1.4.4.

Columns of Tables 1.4.5 and 1.4.7 are fourth eigenvectors of a mistuned system for different values of $r$ which define the mistuning vector **p**, Equation (1.4.8).

**Table 1.4.3.** Differential Eigenvectors ($r = 0$) corresponding (Sinha, 2006a) to $\omega_4$ and $\omega_5$ (from Eq. (1.3.11))

| | | |
|---|---|---|
| $x_1$ | 4.1861 | 0.1425 |
| $x_2$ | 1.1580 | 4.0253 |
| $x_3$ | −3.4704 | 2.3452 |
| $x_4$ | −3.3029 | −2.5758 |
| $x_5$ | 1.4291 | −3.9372 |
| $x_6$ | 4.1861 | 0.1425 |
| $x_7$ | 1.1580 | 4.0253 |
| $x_8$ | −3.4704 | 2.3452 |
| $x_9$ | −3.3029 | −2.5758 |
| $x_{10}$ | 1.4291 | −3.9372 |

**Table 1.4.4.** Elements of an unit mistuning vector (Sinha, 2006a)

| $\chi_{11}$ | $\chi_{12}$ | $\chi_{13}$ | $\chi_{14}$ | $\chi_{15}$ | $\chi_{16}$ | $\chi_{17}$ | $\chi_{18}$ | $\chi_{19}$ | $\chi_{110}$ |
|---|---|---|---|---|---|---|---|---|---|
| −0.1252 | −0.3195 | 0.1885 | 0.5637 | −0.2069 | −0.3207 | −0.0933 | 0.2511 | −0.3235 | −0.4549 |

**Table 1.4.5.** Fourth eigenvector (Sinha, 2006a) of a mistuned system ($\bar{r} = r / 2.6407$)

| $\bar{r} =$ | −1e4 | −9e3 | −8e3 | −7e3 | −6e3 | −5e3 | −4e3 | −3e3 | −2e3 | −1e3 |
|---|---|---|---|---|---|---|---|---|---|---|
| $x_1$ | 4.1299 | 4.1365 | 4.1430 | 4.1493 | 4.1555 | 4.1615 | 4.1671 | 4.1725 | 4.1774 | 4.1820 |
| $x_2$ | 1.3420 | 1.3306 | 1.3178 | 1.3035 | 1.2877 | 1.2703 | 1.2512 | 1.2305 | 1.2081 | 1.1839 |
| $x_3$ | −3.0783 | −3.1148 | −3.1522 | −3.1903 | −3.2291 | −3.2685 | −3.3084 | −3.3487 | −3.3892 | −3.4298 |
| $x_4$ | −2.8453 | −2.8946 | −2.9432 | −2.9913 | −3.0387 | −3.0853 | −3.1310 | −3.1757 | −3.2193 | −3.2617 |
| $x_5$ | 2.3092 | 2.2296 | 2.1480 | 2.0644 | 1.9790 | 1.8917 | 1.8026 | 1.7118 | 1.6192 | 1.5250 |
| $x_6$ | 4.5039 | 4.4753 | 4.4458 | 4.4156 | 4.3847 | 4.3531 | 4.3209 | 4.2880 | 4.2546 | 4.2206 |
| $x_7$ | 1.2233 | 1.2095 | 1.1971 | 1.1864 | 1.1772 | 1.1697 | 1.1639 | 1.1598 | 1.1574 | 1.1568 |
| $x_8$ | −3.7061 | −3.6897 | −3.6716 | −3.6520 | −3.6307 | −3.6078 | −3.5834 | −3.5574 | −3.5299 | −3.5009 |
| $x_9$ | −2.8985 | −2.9413 | −2.9836 | −3.0254 | −3.0667 | −3.1075 | −3.1477 | −3.1873 | −3.2264 | −3.2649 |
| $x_{10}$ | 1.4280 | 1.4251 | 1.4229 | 1.4212 | 1.4202 | 1.4199 | 1.4203 | 1.4214 | 1.4232 | 1.4258 |

All these eigenvectors are close to the first differentiable eigenvector in Table 1.4.3. Tables 1.4.6 and 1.4.8 contain fourth eigenvectors predicted by linearization, that is, after neglecting second and higher derivatives in Equation (1.4.15). The derivative of the eigenvector has been computed from Equation (1.3.13). Comparing columns in Tables 1.4.5 and 1.4.6 and in Tables 1.4.7 and 1.4.8, the linearized analysis is found to yield fairly accurate results. Natural frequencies $\omega_4$ and $\omega_5$are plotted as a

**Table 1.4.6.** Fourth eigenvector (Sinha, 2006a) of a mistuned system using linearization ($\bar{r} = r / 2.6407$)

| $\bar{r} =$ | −1e4 | −9e3 | −8e3 | −7e3 | −6e3 | −5e3 | −4e3 | −3e3 | −2e3 | −1e3 |
|---|---|---|---|---|---|---|---|---|---|---|
| $x_1$ | 4.1474 | 4.1513 | 4.1551 | 4.1590 | 4.1629 | 4.1667 | 4.1706 | 4.1745 | 4.1784 | 4.1822 |
| $x_2$ | 1.4260 | 1.3992 | 1.3724 | 1.3456 | 1.3188 | 1.2920 | 1.2652 | 1.2384 | 1.2116 | 1.1848 |
| $x_3$ | −3.0648 | −3.1054 | −3.1460 | −3.1865 | −3.2271 | −3.2676 | −3.3082 | −3.3487 | −3.3893 | −3.4299 |
| $x_4$ | −2.8985 | −2.9389 | −2.9794 | −3.0198 | −3.0602 | −3.1007 | −3.1411 | −3.1815 | −3.2220 | −3.2624 |
| $x_5$ | 2.3951 | 2.2985 | 2.2019 | 2.1053 | 2.0087 | 1.9121 | 1.8155 | 1.7189 | 1.6223 | 1.5257 |
| $x_6$ | 4.5336 | 4.4988 | 4.4641 | 4.4293 | 4.3946 | 4.3598 | 4.3251 | 4.2904 | 4.2556 | 4.2209 |
| $x_7$ | 1.1370 | 1.1391 | 1.1412 | 1.1433 | 1.1454 | 1.1475 | 1.1496 | 1.1517 | 1.1538 | 1.1559 |
| $x_8$ | −3.7819 | −3.7508 | −3.7196 | −3.6885 | −3.6573 | −3.6262 | −3.5950 | −3.5639 | −3.5327 | −3.5016 |
| $x_9$ | −2.9264 | −2.9640 | −3.0017 | −3.0393 | −3.0770 | −3.1146 | −3.1523 | −3.1899 | −3.2276 | −3.2652 |
| $x_{10}$ | 1.3918 | 1.3955 | 1.3993 | 1.4030 | 1.4067 | 1.4105 | 1.4142 | 1.4179 | 1.4217 | 1.4254 |

**Table 1.4.7.** Fourth eigenvector (Sinha, 2006a) of a mistuned system ($\bar{r} = r / 2.6407$)

| $\bar{r} =$ | 1e3 | 2e3 | 3e3 | 4e3 | 5e3 | 6e3 | 7e3 | 8e3 | 9e3 | 1e4 |
|---|---|---|---|---|---|---|---|---|---|---|
| $x_1$ | 4.1897 | 4.1928 | 4.1954 | 4.1973 | 4.1986 | 4.1993 | 4.1993 | 4.1985 | 4.1971 | 4.1949 |
| $x_2$ | 1.1303 | 1.1009 | 1.0696 | 1.0365 | 1.0017 | 0.9651 | 0.9267 | 0.8867 | 0.8450 | 0.8017 |
| $x_3$ | −3.5109 | −3.5510 | −3.5908 | −3.6299 | −3.6683 | −3.7059 | −3.7423 | −3.7776 | −3.8116 | −3.8440 |
| $x_4$ | −3.3426 | −3.3808 | −3.4173 | −3.4522 | −3.4852 | −3.5163 | −3.5453 | −3.5722 | −3.5970 | −3.6195 |
| $x_5$ | 1.3318 | 1.2330 | 1.1329 | 1.0314 | 0.9288 | 0.8251 | 0.7205 | 0.6149 | 0.5085 | 0.4015 |
| $x_6$ | 4.1511 | 4.1157 | 4.0799 | 4.0437 | 4.0073 | 3.9705 | 3.9335 | 3.8963 | 3.8589 | 3.8215 |
| $x_7$ | 1.1610 | 1.1658 | 1.1725 | 1.1809 | 1.1911 | 1.2032 | 1.2170 | 1.2325 | 1.2498 | 1.2688 |
| $x_8$ | −3.4386 | −3.4054 | −3.3709 | −3.3352 | −3.2984 | −3.2605 | −3.2215 | −3.1816 | −3.1408 | −3.0992 |
| $x_9$ | −3.3402 | −3.3770 | −3.4133 | −3.4490 | −3.4842 | −3.5189 | −3.5531 | −3.5869 | −3.6203 | −3.6534 |
| $x_{10}$ | 1.4333 | 1.4382 | 1.4440 | 1.4506 | 1.4580 | 1.4663 | 1.4755 | 1.4856 | 1.4966 | 1.5084 |

**Table 1.4.8.** Fourth eigenvector (Sinha, 2006a) of a mistuned system using linearization ($\bar{r} = r / 2.6407$)

| $\bar{r} =$ | 1e3 | 2e3 | 3e3 | 4e3 | 5e3 | 6e3 | 7e3 | 8e3 | 9e3 | 1e4 |
|---|---|---|---|---|---|---|---|---|---|---|
| $x_1$ | 4.1900 | 4.1939 | 4.1977 | 4.2016 | 4.2055 | 4.2094 | 4.2132 | 4.2171 | 4.2210 | 4.2248 |
| $x_2$ | 1.1312 | 1.1044 | 1.0776 | 1.0508 | 1.0240 | 0.9972 | 0.9704 | 0.9436 | 0.9168 | 0.8900 |
| $x_3$ | −3.5110 | −3.5515 | −3.5921 | −3.6326 | −3.6732 | −3.7138 | −3.7543 | −3.7949 | −3.8354 | −3.8760 |
| $x_4$ | −3.3433 | −3.3837 | −3.4242 | −3.4646 | −3.5050 | −3.5455 | −3.5859 | −3.6264 | −3.6668 | −3.7072 |
| $x_5$ | 1.3325 | 1.2359 | 1.1393 | 1.0427 | 0.9461 | 0.8495 | 0.7529 | 0.6563 | 0.5597 | 0.4631 |
| $x_6$ | 4.1514 | 4.1166 | 4.0819 | 4.0471 | 4.0124 | 3.9776 | 3.9429 | 3.9081 | 3.8734 | 3.8387 |
| $x_7$ | 1.1601 | 1.1622 | 1.1643 | 1.1664 | 1.1685 | 1.1706 | 1.1728 | 1.1749 | 1.1770 | 1.1791 |
| $x_8$ | −3.4393 | −3.4081 | −3.3770 | −3.3458 | −3.3147 | −3.2835 | −3.2524 | −3.2212 | −3.1901 | −3.1589 |
| $x_9$ | −3.3405 | −3.3782 | −3.4158 | −3.4535 | −3.4911 | −3.5287 | −3.5664 | −3.6040 | −3.6417 | −3.6793 |
| $x_{10}$ | 1.4329 | 1.4366 | 1.4403 | 1.4441 | 1.4478 | 1.4515 | 1.4553 | 1.4590 | 1.4628 | 1.4665 |

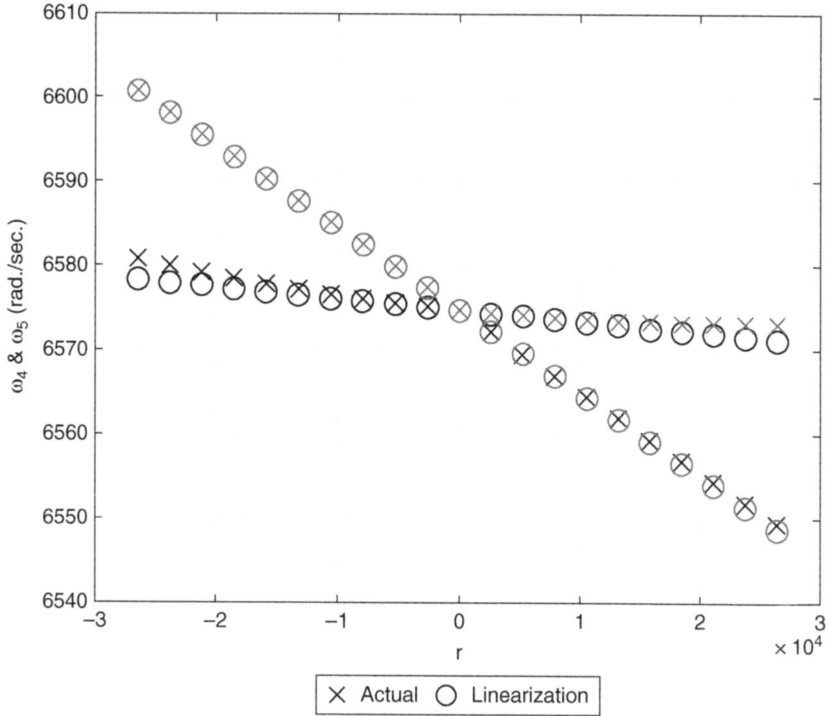



**Figure 1.4.4.** Prediction of $\omega_4$ and $\omega_5$ (Sinha, 2006a) using linearization (range of $r$ axis: −3e4 to +3e4).

function of $r$ in Figure 1.4.4. Once again, the linearized analysis yields fairly good prediction of natural frequencies. The first derivative in Equation (1.4.14) has been obtained through Equation (1.3.11).

In Figure 1.4.5, the changes in frequencies due to mistuning are shown for those values of coupling stiffnesses for which tuned frequencies are shown in Figure 1.2.3. Frequencies pair (2,3), (4,5), (6,7), and (8,9) correspond to repeated frequencies numbers 1, 2, 3, and 4. Mistuned modes are represented as coefficients of tuned modes:

$$\mathbf{v}_i = \mathbf{\Phi}\mathbf{y}_i \tag{1.4.18}$$

In Figure 1.4.6, coefficients $\mathbf{y}_i$ of the fourth mistuned mode are presented for values of coupling stiffnesses same as those in Figure 1.4.5. For larger coupling stiffnesses $\chi = 0.325$ and $\chi = 0.6$, dominant coefficients correspond to third and ninth tuned modes. For small coupling stiffness $\chi = 0.05$, there are significantly more contributions from modes other than third and ninth modes. This is because, all eigenvalues are quite close together, and as shown by Equation (1.3.34), and the contributions from other modes in change in an eigenvector are significantly enhanced.

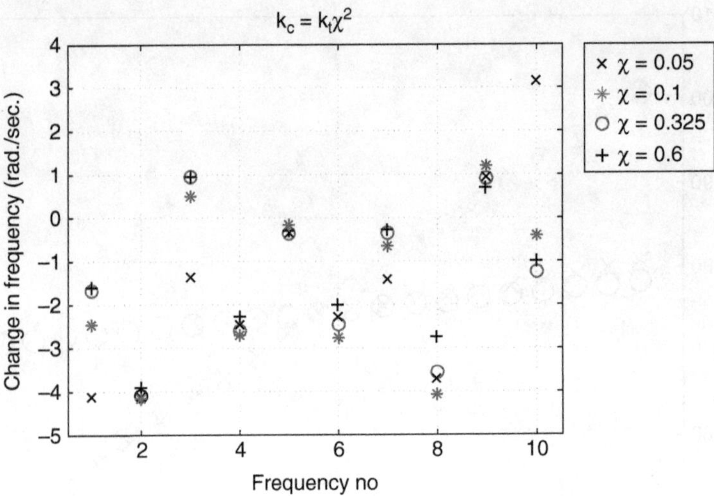

**Figure 1.4.5.**    Change in frequency due to mistuning ($\bar{r} = 1000$) in Table 1.4.7 ($k_t = 430000\,N/m$, $m_t = 0.0114\,kg$).

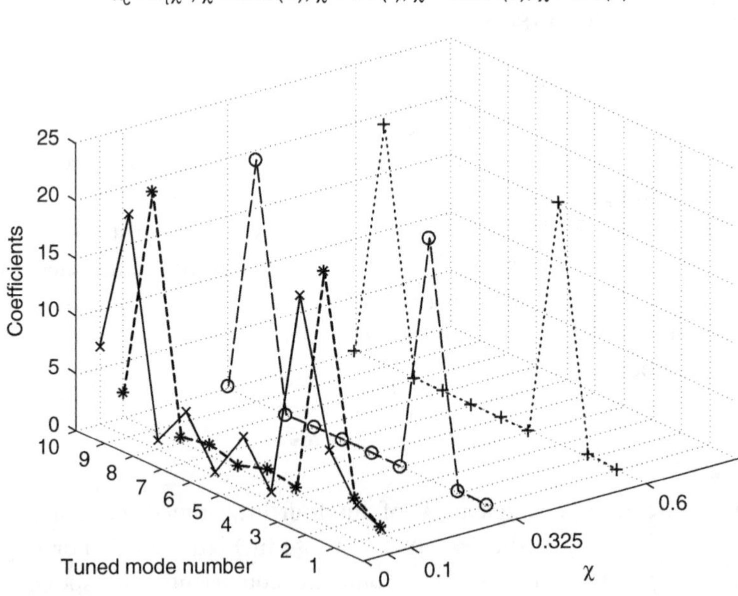

**Figure 1.4.6.**    Components of the fourth mistuned mode, mistuning ($\bar{r} = 1000$) in Table 1.4.7 ($k_t = 430000\,N/m$, $m_t = 0.0114\,kg$).

## 1.5 Special Treatment of Extremely Low Coupling Stiffness

Assuming that $\Delta M = 0$, Equation (1.3.2) is rewritten as

$$(K_d + \Delta K_d)\mathbf{v} = \lambda M_t \mathbf{v} \tag{1.5.1}$$

where the diagonal matrix $K_d$ is carved out of the matrix $K$, Equation (1.1.6), by eliminating $k_c$, that is,

$$K_d = \begin{bmatrix} k_t + \delta k_1 & & & & \\ & k_t + \delta k_2 & & & \\ & & \ddots & & \\ & & & k_t + \delta k_{n-1} & \\ & & & & k_t + \delta k_n \end{bmatrix} \tag{1.5.2}$$

and

$$\Delta K_d = \begin{bmatrix} 2k_c & -k_c & 0 & . & . & -k_c \\ -k_c & 2k_c & -k_c & 0 & . & 0 \\ 0 & -k_c & 2k_c & -k_c & - & 0 \\ . & . & . & . & . & . \\ -k_c & 0 & . & 0 & -k_c & 2k_c \end{bmatrix} \tag{1.5.3}$$

Here, when $\Delta K_d = 0$,

$$\lambda_i = \frac{k_t + \delta k_i}{m_t}; i = 1, 2, \ldots, n \tag{1.5.4}$$

and

$$\mathbf{v}_i = \frac{1}{\sqrt{m_t}} \begin{bmatrix} 0 & \cdots & 0 & 1 & 0 & \cdots & 0 \end{bmatrix}^T = \frac{1}{\sqrt{m_t}} \mathbf{e}_i \tag{1.5.5}$$

where $\mathbf{e}_i$ is the ith column of the identity matrix. Treating $\Delta K_d$ as a perturbation in the stiffness matrix, Equation (1.3.26) yields

$$\delta\lambda_i = \mathbf{v}_i^T \Delta K_d \mathbf{v}_i = \frac{2k_c}{m_t} \tag{1.5.6}$$

From Equations (1.3.27) – (1.3.31),

$$\delta\mathbf{v}_i = P^{-1}\mathbf{q} \tag{1.5.7}$$

where

$$P = \begin{bmatrix} \mathbf{v}_1 & . & \mathbf{v}_{i-1} & 2\mathbf{v}_i & \mathbf{v}_{i+1} & . & \mathbf{v}_{nd} \end{bmatrix}^T m_t \tag{1.5.8}$$

and the jth element of the vector **q** is given by

$$q_j = \frac{-\mathbf{v}_j^T \delta K_d \mathbf{v}_i}{(\lambda_j - \lambda_i)}; \ j \neq i \tag{1.5.9}$$

Equation (1.3.27) yields

$$q_{i+1} = \frac{-k_c}{m_t(\lambda_{i+1} - \lambda_i)} = \frac{-k_c}{\delta k_{i+1} - \delta k_i} \tag{1.5.10}$$

$$q_{i-1} = \frac{-k_c}{m_t(\lambda_{i-1} - \lambda_i)} = \frac{-k_c}{\delta k_{i-1} - \delta k_i} \tag{1.5.11}$$

Equation (1.3.28) yields

$$q_i = 0 \tag{1.5.12}$$

Equation (1.5.7) yields

$$\delta v_{i,i-1} = -\frac{k_c}{\sqrt{m_t}(\delta k_{i-1} - \delta k_i)} \tag{1.5.13}$$

$$\delta v_{i,i} = 0 \tag{1.5.14}$$

$$\delta v_{i,i+1} = -\frac{k_c}{\sqrt{m_t}(\delta k_{i+1} - \delta k_i)} \tag{1.5.15}$$

This result shows that mistuned modes will be highly localized (Wei and Pierre, 1988), which is indicated by participation of all modes in Figure 1.4.6 for a low coupling stiffness $\chi = 0.05$.


## 1.6    Frequency Veering Phenomenon

Tuned bladed disks or periodic structures exhibit frequency veering phenomenon, which refers to natural frequencies of a tuned system coming closer and then veering away from each other as the nodal diameter or harmonic index increases. This veering region has been associated with the mode localization and amplification of forced response amplitude because of mistuning (Kenyon, Griffin, and Kim, 2005).

To visualize frequency veering phenomenon, consider the model in Figure 1.6.1, which has two masses per sector. The spring with stiffness $k_i$ and mass $m_i$ represent blade's modal stiffness and modal mass, respectively. The spring with stiffness $k_g$ and mass $m_g$ represent disk's modal stiffness and modal mass, respectively. The stiffness $k_c$ simulates the structural coupling between blades in circumferential direction.

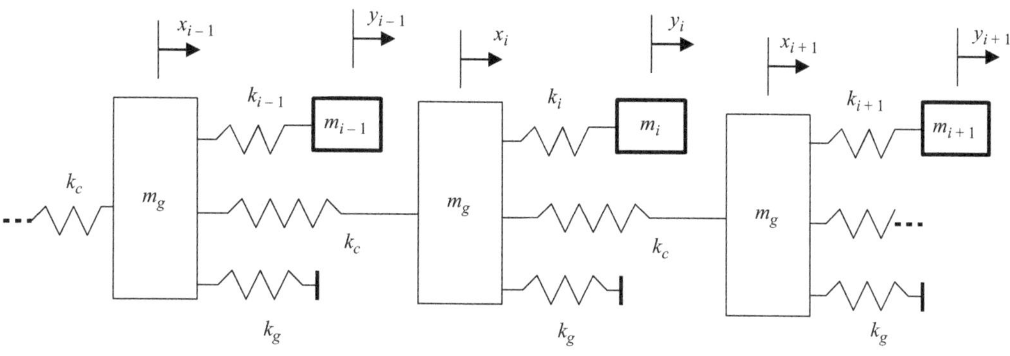

**Figure 1.6.1.** Two degrees of freedom per sector model.

The differential equations of motion are

$$m_g \ddot{x}_i + (k_g + 2k_c + k_i)x_i - k_i y_i - k_c x_{i-1} - k_c x_{i+1} = 0; \quad i = 1, 2, \dots, n \quad (1.6.1)$$

$$m_i \ddot{y}_i - k_i x_i + k_i y_i = 0 \quad (1.6.2)$$

Also, $i+1 = 1$ when $i = n$, and $i-1 = n$ when $i = 1$. In matrix form,

$$M\ddot{\mathbf{z}} + K\mathbf{z} = 0 \quad (1.6.3a)$$

$$\mathbf{z}^T = \begin{bmatrix} x_1 & y_1 & x_2 & y_2 & \cdots & x_n & y_n \end{bmatrix} \quad (1.6.3b)$$

where

$$M = \begin{bmatrix} M_1 & & & & \\ & M_2 & & & \\ & & M_3 & & \\ & & & \ddots & \\ & & & & M_n \end{bmatrix} \quad (1.6.4)$$

$$M_i = \begin{bmatrix} m_g & 0 \\ 0 & m_i \end{bmatrix}; \quad i = 1, 2, \cdots, n \quad (1.6.5)$$

$$K = \begin{bmatrix} K_1 & -K_c & 0 & \cdots & -K_c \\ -K_c & K_2 & -K_c & \cdots & 0 \\ . & . & . & . & . \\ . & . & . & . & . \\ -K_c & 0 & \cdots & -K_c & -K_n \end{bmatrix} \quad (1.6.6)$$

$$K_i = \begin{bmatrix} k_g + 2k_c + k_i & -k_i \\ -k_i & k_i \end{bmatrix}; \quad i = 1, 2, \cdots, n \quad (1.6.7)$$

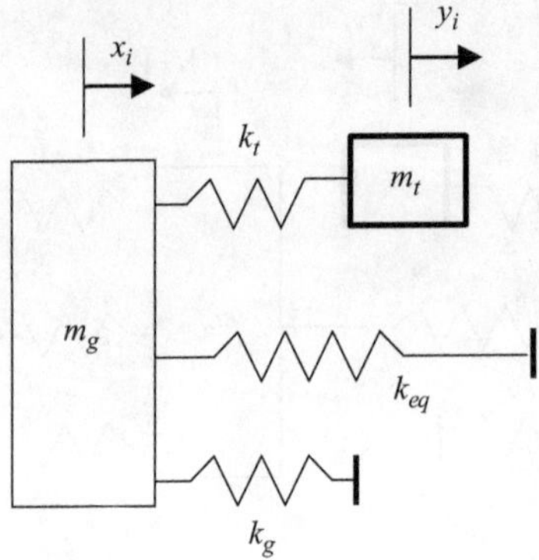


**Figure 1.6.2.**   Equivalent sector model for the tuned system.

$$K_c = \begin{bmatrix} k_c & 0 \\ 0 & 0 \end{bmatrix} \tag{1.6.8}$$

An equivalent single sector (two degree of freedom) of model, Figure 1.6.2, can be obtained for each harmonic index $\ell$ as the tuned modes satisfy (Equation 1.2.21)

$$x_{i+1} = e^{j\phi\ell} x_i \quad \text{and} \quad x_{i-1} = e^{-j\phi\ell} x_i \tag{1.6.9}$$

Substituting Equation (1.6.9) into Equation (1.6.1) for the tuned system,

$$m_g \ddot{x}_i + (k_{ge} + k_t)x_i - k_i y_i = 0 \tag{1.6.10}$$

where

$$k_{ge} = k_g + k_{eq}; \; k_{eq} = 4k_c \sin^2\left(\frac{\phi\ell}{2}\right) \tag{1.6.11}$$

Combining Equations (1.6.10) and (1.6.2), equivalent stiffness and mass matrices are

$$K_s = \begin{bmatrix} k_{ge} + k_t & -k_t \\ -k_t & k_t \end{bmatrix} \quad \text{and} \quad M_s = \begin{bmatrix} m_g & 0 \\ 0 & m_t \end{bmatrix} \tag{1.6.12}$$

Natural frequencies are given by the roots of the following equations:

$$\det(K_s - \omega^2 M_s) = m_t m_g \omega^4 - [(k_{ge} + k_t)m_t + k_t m_g]\omega^2 + k_{ge}k_t = 0 \tag{1.6.13}$$

Natural frequencies are plotted in Figure 1.6.3 as a function of nodal diameters $\ell = 0,1,2,\cdots,8$ for a sixteen-bladed tuned rotor. Parameter values (Kenyon et al., 2005) are (in consistent units): $m_t = 15$, $m_g = 250$, $k_t = 23000$, $k_g = 10000$ and $k_c = 600,000$. For every nodal diameter, there are two frequencies, lower and

**Table 1.6.1.** Modal vectors for higher frequencies

| ND | 0 | 1 | 2 | 3 | 4 | 5 | 6 | 7 | 8 |
|----|----|----|----|----|----|----|----|----|----|
| xd | −0.0154 | −0.0195 | −0.0449 | −0.0617 | −0.0629 | −0.0631 | −0.0631 | −0.0632 | −0.0632 |
| xb | 0.2504 | 0.2456 | 0.1820 | 0.0574 | 0.0280 | 0.0184 | 0.0142 | 0.0123 | 0.0118 |

**Table 1.6.2.** Modal vectors for lower frequencies

| ND | 0 | 1 | 2 | 3 | 4 | 5 | 6 | 7 | 8 |
|----|----|----|----|----|----|----|----|----|----|
| xd | −0.0613 | −0.0602 | −0.0446 | −0.0141 | −0.0069 | −0.0045 | −0.0035 | −0.0030 | −0.0029 |
| xb | −0.0629 | −0.0797 | −0.1831 | −0.2517 | −0.2567 | −0.2575 | −0.2578 | −0.2579 | −0.2579 |

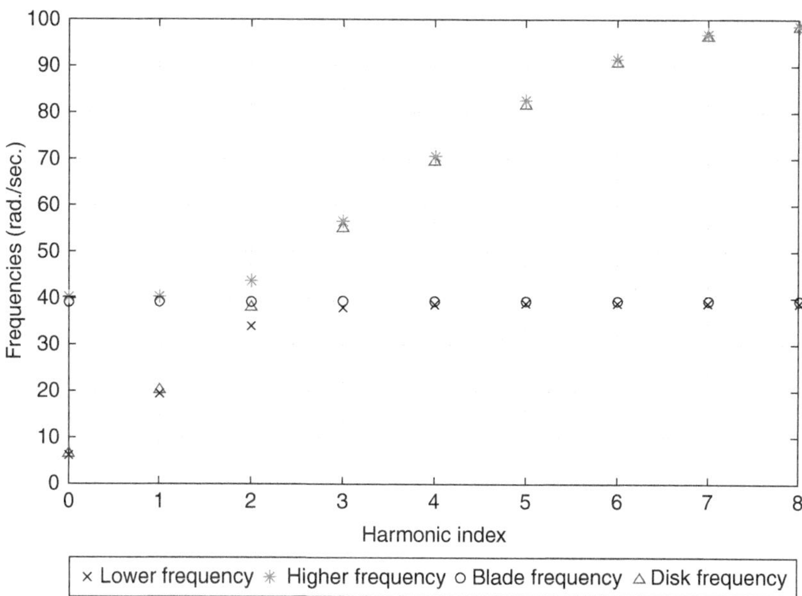


**Figure 1.6.3.** Natural frequencies of a perfectly tuned two degree of freedom per sector model.

higher. Corresponding modal vectors are shown in Tables 1.6.1 and 1.6.2. Blade alone frequencies

$$\omega_b = \sqrt{\frac{k_t}{m_t}} \tag{1.6.14}$$

are also plotted. Disk alone frequencies

$$\omega_g = \sqrt{\frac{k_g + 4k_c \sin^2(\phi/2)}{m_g}} \tag{1.6.15}$$

are also plotted in Figure 1.6.3.

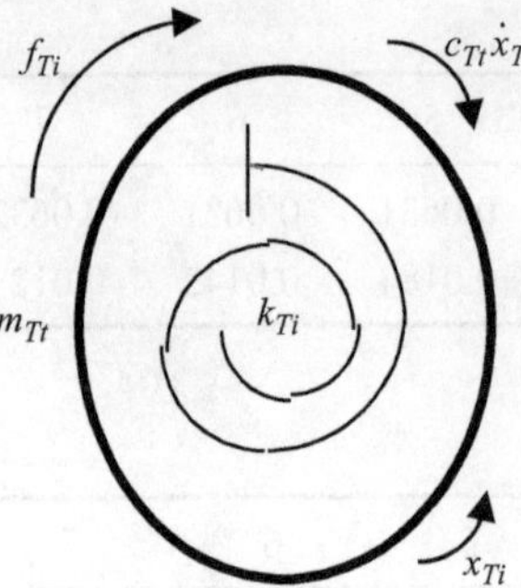



**Figure 1.7.1.**   A blade in torsional mode of vibration.

For nodal diameter 0 and 1, lower and higher frequencies are close to those of disk alone and blade alone frequencies, Figure 1.6.3. At nodal diameter = 2, both frequencies of bladed disk are closest. After nodal diameter = 2, both frequencies veer away from each other. Further, higher and lower frequencies become closer to those of disk alone and blade alone frequencies, respectively.

## 1.7    Flutter and Mistuning

Flutter or aeroelastic instability (Dowell, 2014) occurs when work done by aerodynamic forces exceeds the total energy dissipation in the system. Kaza and Kielb (1982), Kielb and Kaza (1983), and Crawley and Hall (1985) have shown that mistuning has stabilizing effects on the rotor.

Consider the system shown in Figure 1.7.1, where each blade is represented by its torsional mode of vibration. The modal mass moment of inertia and modal torsional stiffness of blade number $i$ are $m_{Ti}$ and $k_{Ti}$, respectively. Mistuning is again represented by variations in stiffness alone, that is,

$$k_{Ti} = k_{Tt}(1 + \delta k_i) \tag{1.7.1}$$

where $k_{Ti}$ is the torsional stiffness of the tuned system, and $\delta k_i$ represents deviations in torsional stiffnesses due to mistuning. The damping coefficient from structural sources is $c_{Tt}$.

Differential equation of motion for each blade is

$$m_{Ti}\ddot{x}_{Ti} + c_{Ti}\dot{x}_{Ti} + k_{Ti}x_{Ti} = f_{Ti}^{a}(\mathbf{x}_T, \dot{\mathbf{x}}_T) \tag{1.7.2}$$

where the aerodynamic torque on each blade $f_{Ti}^{a}$ depends on positions and velocities of all blades and

$$\mathbf{x}_T = [x_{T1} \quad x_{T2} \quad \cdots \quad x_{Tn}]^T \tag{1.7.3}$$

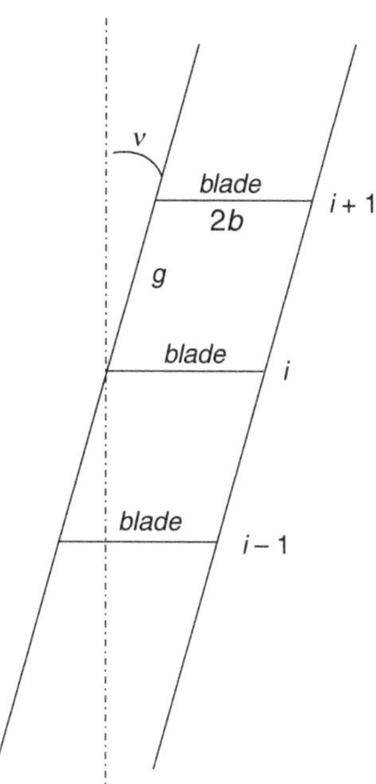



**Figure 1.7.2.**  Description of cascade geometry.

The aerodynamic torque on each blade $f_{Ti}^a$ is computed (Kaza and Kielb, 1982) as

$$f_{Ti}^a = \sum_{r=0}^{n-1} f_{Tri}^a \qquad (1.7.4)$$

where $f_{Tri}^a$ is the moment acting on the blade number $i$ vibrating sinusoidally in tuned mode number $r$ with frequency $\omega$ and amplitude $\beta_r$; and is commonly described by using the cascade model (Figure 1.7.2) of a rotor stage as follows:

$$f_{Tri}^a = \pi \rho b^4 \omega^2 \ell_{\alpha\alpha r} \beta_r e^{(j\omega t + (i-1)\phi_r)} \qquad (1.7.5)$$

where $\rho$ and $2b$ are fluid density and blade's chord length, respectively. The inter-blade phase angle $\phi_r$ for the tuned mode $r$ is:

$$\phi_r = \frac{2\pi r}{n}; \; r = 0,1,\ldots,n-1 \qquad (1.7.6)$$

The coefficient $\ell_{\alpha\alpha r}$ is computed as

$$\ell_{\alpha\alpha r} = \frac{4}{\kappa^2} c_{T\alpha r} \qquad (1.7.7)$$

where aerodynamic coefficient $c_{T\alpha r}$ for each tuned mode is computed using codes for incompressible or compressible flow, for example, codes by Whitehead (1960) and Adamczyk and Goldstein (1978). Input parameters for these codes are gap to chord ratio ($g/2b$), stagger angle ($v$), interblade phase angle ($\phi_r$) and reduced frequency $\kappa$ defined for fluid velocity $V$ as

$$\kappa = \frac{\omega b}{V} \tag{1.7.8}$$

From Equations (1.7.2), (1.7.5), and (1.7.6),

$$\ddot{x}_{Ti} + 2\xi_s \frac{\omega_{nt}^2}{\omega}\dot{x}_{Ti} + \omega_{nt}^2(1+\delta k_i) = \frac{1}{\mu\gamma^2}\omega^2 \sum_{r=0}^{n-1} \ell_{\alpha\alpha r}\beta_r e^{j(\omega t+(i-1)\phi_r)} \tag{1.7.9}$$

$$\gamma^2 = \frac{m_{Tt}}{m_t b^2}; \quad \omega_{nt}^2 = \frac{k_{Tt}}{m_{Tt}} \tag{1.7.10a, b}$$

$$\xi_s = \frac{c_t\omega}{2m_{Tt}\omega_{nt}^2}; \quad \mu = \frac{m_t}{\pi\rho b^2} \tag{1.7.11a, b}$$

Let

$$x_{Ti} = a_{Ti}e^{j\omega t}; \quad j = \sqrt{-1} \tag{1.7.12}$$

Substituting Equation (1.7.12) into Equation (1.7.9) and equating the coefficient of $e^{j\omega t}$ on both sides,

$$-\lambda^2 \mathbf{a}_{Ti} + (1+2j\xi_s + \delta k_i)a_{Ti} = \frac{\lambda^2}{\mu\gamma^2}\sum_{r=0}^{n-1}\ell_{\alpha\alpha r}\beta_r e^{j(i-1)\phi_r} \tag{1.7.13}$$

where

$$\lambda = \frac{\omega}{\omega_{nt}} \tag{1.7.14}$$

Equation (1.7.13) can be written as

$$-\lambda^2 \mathbf{a}_T + (1+2j\xi_s)\mathbf{a}_T + \Lambda\mathbf{a}_T = \frac{\lambda^2}{\mu\gamma^2}EL\boldsymbol{\beta} \tag{1.7.15}$$

where

$$\mathbf{a}_T = \begin{bmatrix} a_{T1} \\ a_{T2} \\ \vdots \\ a_{Tn} \end{bmatrix}, \quad \boldsymbol{\beta} = \begin{bmatrix} \beta_1 \\ \beta_2 \\ \vdots \\ \beta_n \end{bmatrix} \tag{1.7.16a, b}$$

$$
L = \begin{bmatrix} \ell_{\alpha\alpha 1} & & & \\ & \ell_{\alpha\alpha 2} & & \\ & & \ddots & \\ & & & \ell_{\alpha\alpha n} \end{bmatrix}; \quad \Lambda = \begin{bmatrix} \delta k_1 & & & \\ & \delta k_2 & & \\ & & \ddots & \\ & & & \delta k_n \end{bmatrix} \quad (1.7.17a, b)
$$

and

$$
E = \begin{bmatrix} 1 & 1 & \cdots & 1 & 1 \\ 1 & e^{j\phi_1} & \cdots & e^{j\phi_{n-2}} & e^{j\phi_{n-1}} \\ 1 & e^{j2\phi_1} & \cdots & e^{j2\phi_{n-2}} & e^{j2\phi_{n-1}} \\ \vdots & \vdots & \vdots & \vdots & \vdots \\ 1 & e^{j(n-1)\phi_1} & \cdots & e^{j(n-1)\phi_{n-2}} & e^{j(n-1)\phi_{n-1}} \end{bmatrix} \quad (1.7.18)
$$

Further

$$
\mathbf{a}_T = E\boldsymbol{\beta} \quad (1.7.19)
$$

Substituting Equation (1.7.19) into Equation (1.7.15),

$$
P\mathbf{a}_T = \lambda^2 Q\mathbf{a}_T \quad (1.7.20)
$$

where

$$
P = ((1 + 2j\xi_s)I_n + \Lambda) \quad (1.7.21)
$$

and

$$
Q = \left( \frac{1}{\mu\gamma^2} ELE^{-1} + I_n \right) \quad (1.7.22)
$$

Equation (1.7.20) indicates that $\lambda^2$ and $\mathbf{a}_T$ are generalized eigenvalue and eigenvector of complex matrices $P$ and $Q$. Then, eigenvalues of equivalent 2n system of first order differential equations are described as

$$
s = j\lambda = s_R + js_I \quad (1.7.23)
$$

The flutter or aeroelastic instability occurs when $s_R > 0$.

For $\mu = 258$, $\xi_s = 0$, and $\gamma = 0.5774$, eigenvalues $(s)$ are plotted in Figure 1.7.3 for a perfectly tuned system. Some of the eigenvalues have positive real parts and the perfectly tuned system is unstable. Cascade parameters and aerodynamic coefficients are presented in Table 1.7.1.

Next, mistuning factors $\delta k_i$ are generated randomly with standard deviation = 0.03, Figure 1.7.4. Eigenvalues of the mistuned system are shown in Figure 1.7.5. All eigenvalues have negative real parts and the mistuning has been able to stabilize the system.

**Table 1.7.1.** Aerodynamic coefficients $c_{Tar}$ (Whitehead, 1960)
Reduced Frequency $\kappa = 0.2$, Stagger Angle $v = 60^0$, space to chord ratio $g\,/\,(2b) = 1$, $n = 10$

| $r$ | $\phi_r\,/\,(2\pi)$ | $c_{Tar}$ |
|---|---|---|
| 0 | 0 | $-0.1475 - j0.0644$ |
| 1 | 0.1 | $-0.1596 - j0.1381$ |
| 2 | 0.2 | $-0.1997 - j0.1130$ |
| 3 | 0.3 | $-0.2437 - j0.0906$ |
| 4 | 0.4 | $-0.2749 - j0.0705$ |
| 5 | 0.5 | $-0.2852 - j0.0507$ |
| 6 | 0.6 | $-0.2717 - j0.0292$ |
| 7 | 0.7 | $-0.2352 - j0.0049$ |
| 8 | 0.8 | $-0.1792 + j0.0211$ |
| 9 | 0.9 | $-0.1050 + j0.0377$ |

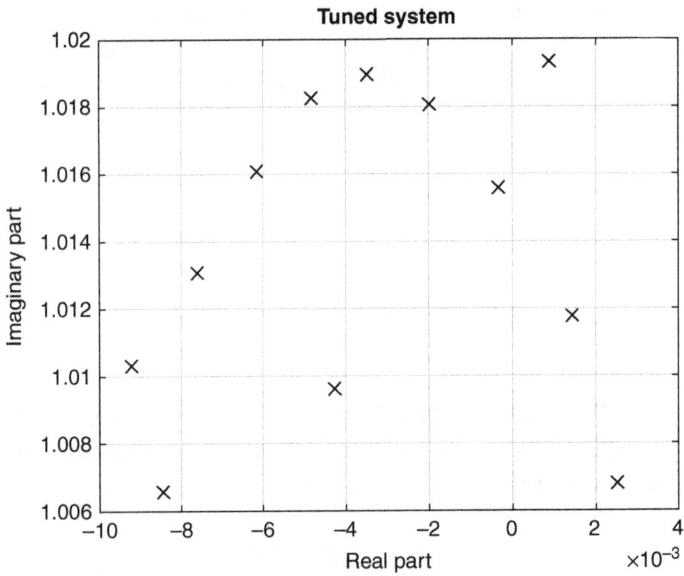


**Figure 1.7.3.**   Eigenvalues of a perfectly tuned system.

## 1.8    Vibration Localization in Atomic Chains

### 1.8.1    Monoatomic Chain

Ideal crystals have periodic structures containing repeated spatial structural units. The simplest periodic structure (Cleland, 2003) is one dimensional and repeated structure contains a single atom as shown in Figure 1.8.1, where $m_t$ is the mass of

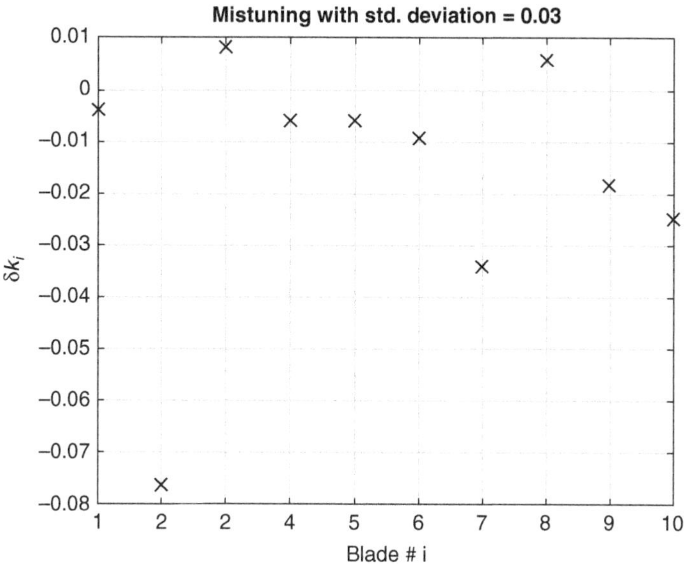



**Figure 1.7.4.** Mistuning factor $\delta k_i$ with standard deviation $= 0.03$.

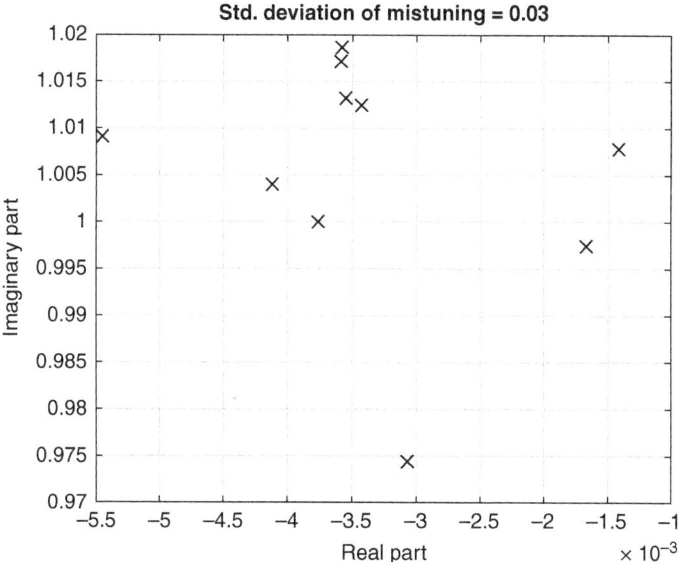



**Figure 1.7.5.** Eigenvalues of a mistuned system.

each atom and the interaction between atoms $i$ and $i+1$ is represented by a linear spring with stiffness $kb_i$. The governing system of differential equations of motion is

$$m_i \ddot{x}_i + kb_i(x_i - x_{i+1}) + kb_{i-1}(x_i - x_{i-1}) = 0;\ i = 1, 2, \ldots, n \qquad (1.8.1)$$

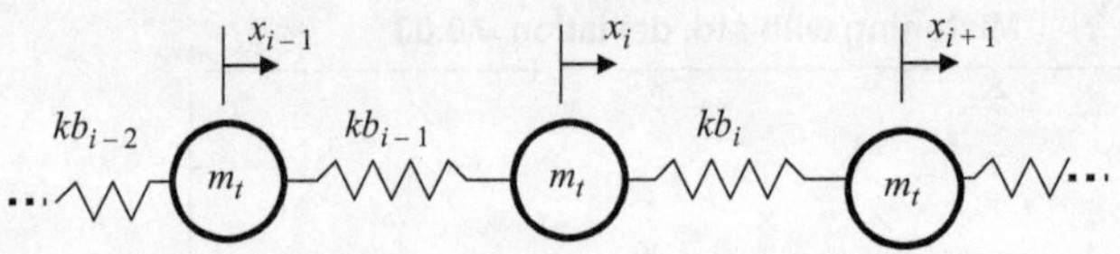

**Figure 1.8.1.**   One-dimensional nearly periodic structure with one atom in each unit.

For $n$ masses in the rotationally periodic structure, $i+1=1$ when $i=n$, and $i-1=n$ when $i=1$.

In matrix form,

$$M\ddot{\mathbf{x}} + K\mathbf{x} = 0 \tag{1.8.2}$$

where

$$\mathbf{x} = \begin{bmatrix} x_1 & x_2 & \cdots & x_{n-1} & x_n \end{bmatrix}^T \tag{1.8.3}$$

$$M = m_t I_n \tag{1.8.4}$$

$$K = \begin{bmatrix} k_{b1} + k_{bn} & -k_{b1} & 0 & \cdots & 0 & -k_{bn} \\ -k_{b1} & k_{b2} + k_{b1} & -k_{b2} & \cdots & 0 & 0 \\ 0 & -k_{b2} & k_{b3} + k_{b2} & \cdots & 0 & 0 \\ \vdots & \vdots & \vdots & \cdots & \vdots & \vdots \\ -k_{bn} & 0 & 0 & \cdots & -k_{bn-1} & k_{bn} + k_{bn-1} \end{bmatrix} \tag{1.8.5}$$

It is interesting to note that this model is quite similar to the basic model of a bladed disk, Figure 1.1.1. The difference is that there is no spring connected to the ground for each atom, Here, the equivalent coupling stiffness $\chi$, Equation (1.2.25), is infinite. Further, disorder or mistuning is introduced to coupling stiffness here.

## Perfectly Tuned System

For a perfectly tuned system, all coupling stiffnesses are identical, that is,

$$kb_1 = kb_2 = \cdots = kb_n = kb_t \tag{1.8.6}$$

Natural frequencies and mode shapes of a perfectly tuned system will be again given by Equations (1.2.4) and (1.2.5). These vibrations are known as phonons (Cleland, 2003; Kittel, 1996). Tuned natural frequencies are

$$\omega_{t\ell} = \sqrt{\dfrac{4kb_t \sin^2\left(\dfrac{\phi\ell}{2}\right)}{m_t}}; \quad \ell = 0,1,2,\ldots,n-1 \tag{1.8.7}$$

From Equation (1.8.7), nondimensional frequencies are defined as

$$\omega_{tt\ell} = 2\left|\sin\left(\frac{\pi\ell}{n}\right)\right| \tag{1.8.8}$$

where

$$\omega_{tt\ell} = \frac{\omega_{t\ell}}{\sqrt{kb_t / m_t}} \tag{1.8.9}$$

and

$$\phi = \frac{2\pi}{n} \tag{1.8.10}$$

Corresponding to constant interatomic phase angle, a wave vector $\kappa$ (Cleland, 2003) is defined as

$$\kappa = \frac{\phi\ell}{a} = \frac{2\pi\ell}{na} \tag{1.8.11a}$$

where $a$ is the interatomic distance at static equilibrium. Using Equations (1.8.8) and (1.8.11),

$$\omega_{tt\ell} = 2\left|\sin\left(\frac{\kappa a}{2}\right)\right| \tag{1.8.11b}$$

For a small value of $\ell$, frequencies $\omega_{tt\ell}$ are almost proportional to the wave vector $\kappa$. But, this is not true for higher values of $\ell$, and the medium through which wave travels is known as dispersive.

Nondimensional frequencies are plotted as a function of interatomic phase angles $\phi\ell$; $\ell = 0,1,2,\cdots,n$, for a 300-atom ring, Figure 1.8.2. The lowest frequency is zero, for which the interatomic phase angle is zero. In this vibratory mode, the system will have pure translational degree of freedom like a rigid body. For interatomic phase angles around 180 degrees, there are many frequencies clustered together. Also, there are repeated frequencies as $\omega_{t\ell} = \omega_{tn-\ell}$. In solid state physics (Kittel, 1996), tuned frequencies are plotted as shown in Figure 1.8.3, where interatomic phase angles lie between $-\pi$ and $\pi$. This range of interatomic phase angles corresponds to the first Brillouin zone, which will be described in section 3.1.2. For a small value of $\ell$, frequencies $\omega_{tt\ell}$ are almost proportional to the wave vector $\kappa$. But, this is not true for higher values of $\ell$, and the medium through which wave travels is known as dispersive (Cleland, 2003), and the frequency plots in Figure 1.8.3 are known as phonon dispersion curves.

### Disorder/Mistuning

Allen and Kelner (1998) studied the effects of random disorder in the stiffness $kb_i$ as follows:

$$kb_i = kb_t(1 + \xi_i); \, i = 1, 2, \dots, n \tag{1.8.12}$$

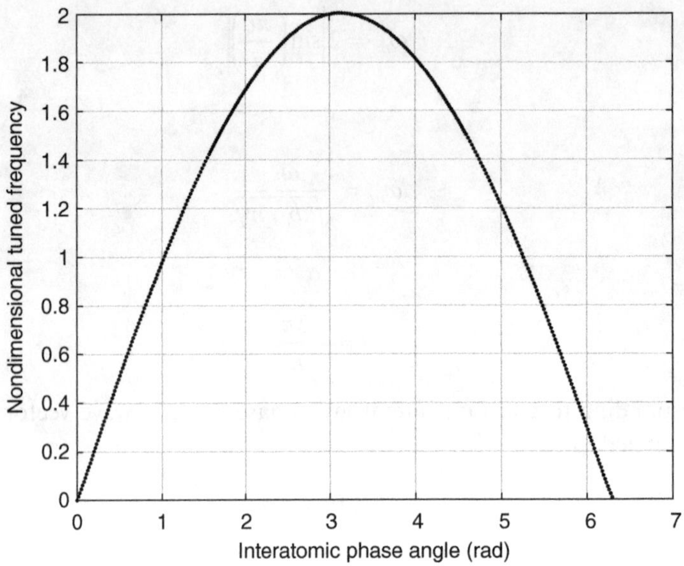

**Figure 1.8.2.**   Tuned natural frequencies for monoatomic chain ($n = 300$).

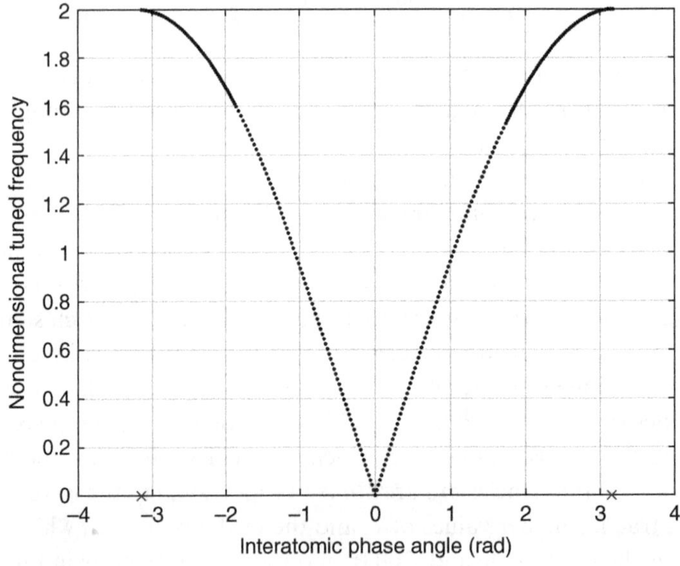

**Figure 1.8.3.**   Tuned natural frequencies for monoatomic chain ($n = 300$) in a Brioullian zone.

where $\xi_i$ are samples from a random variable with zero mean and standard deviation $\sigma$. Natural frequencies (Figure 1.8.4) and corresponding modal vectors are calculated by using the Matlab routine: eig. The standard deviation ($\sigma$) of the normal distribution of stiffness $kb_i$ is selected to be 0.07 as chosen by Allen and Kelner (1999). Now, all repeated frequencies split, and modal vectors are unique.

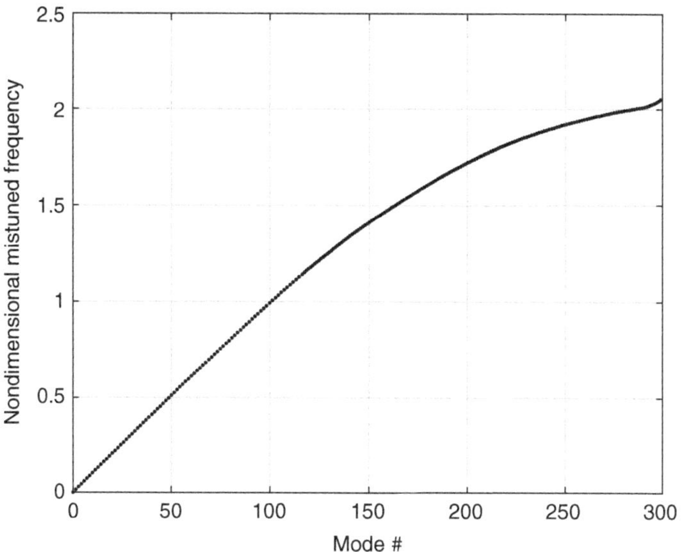

**Figure 1.8.4.** Mistuned natural frequencies for monoatomic chain ($n = 300$, $\sigma = 0.07$).

A participation ratio $p_m(i)$ is defined as a measure of localization for a modal vector $V(:,i)$:

$$p_m(i) = \frac{1}{\displaystyle\sum_{\ell=1}^{n} [|\,V(\ell,i)\,|]^4} \tag{1.8.13}$$

where

$$\sum_{\ell=1}^{n} [|\,V(\ell,i)\,|]^2 = 1 \tag{1.8.14}$$

For a perfectly tuned system,

$$|V(1,i)| = |V(2,i)| = \cdots = |V(n,i)| = \frac{1}{\sqrt{n}} \tag{1.8.15}$$

Substituting Equation (1.8.15) into Equation (1.8.13), $p_m(i) = n$ for each tuned mode. For the most localized mode of vibration, $V(\ell,i) = 0$ for all $\ell$ except one for which it will be unity. In this case, the participation ratio $p_m(i) = 1$. Hence, a value closer to one would imply a higher degree of localization

Participation ratios for all mistuned modes are shown in Figure 1.8.5. The degree of localization for the highest few modes is much higher than those for the first fifty modes, which can be seen in Figures 1.8.6 and 1.8.7. In spite of having an infinite value of the equivalent coupling stiffness $\chi$ (Equation 1.2.25), there are many modes clustered around the interblade phase angle $\pi$, which correspond to highest

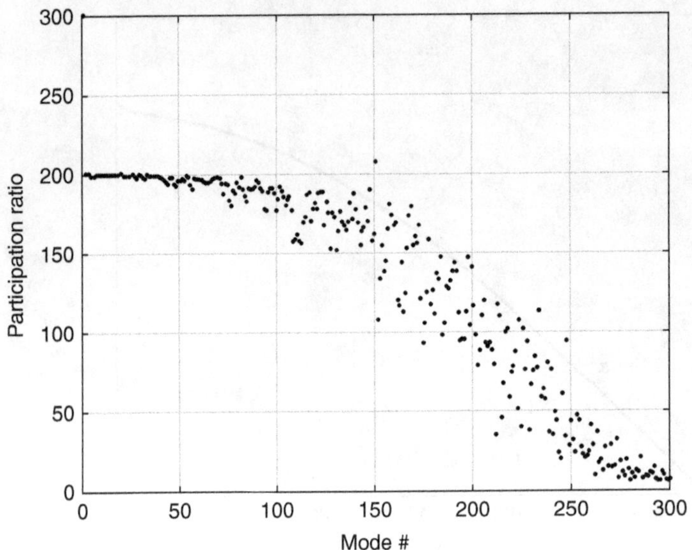

**Figure 1.8.5.**   Participation ratio for mistuned monoatomic chain ($n = 300$, $\sigma = 0.07$).

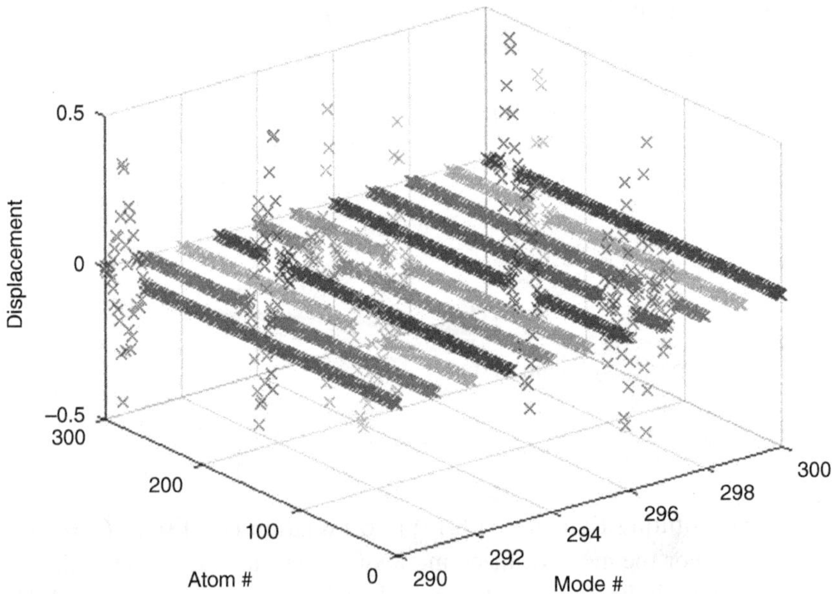

**Figure 1.8.6.**   Localized modal vectors 290–300 of monoatomic chain.

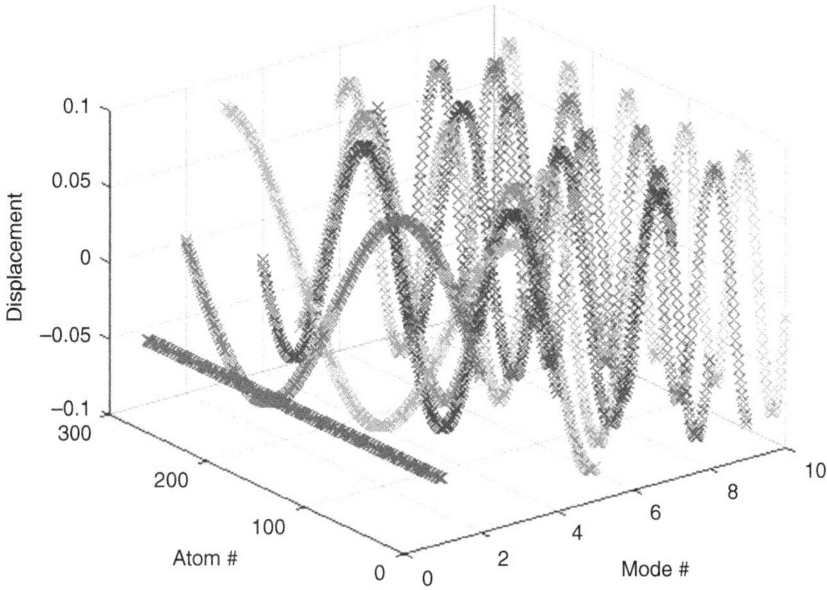


**Figure 1.8.7.**  Nonlocalized modal vectors 1–10 of monoatomic chain.

frequencies. This will imply a greater amount of modal interactions leading to more severe localization of modes, Section 1.5.

## 1.8.2   Diatomic Chain

Consider a one-dimensional periodic structure with two atoms in each unit (Figure 1.8.8) that is being repeated. Kittel (1996) describes this as a model of a cubic crystal where atoms of masses $m_{1t}$ and $m_{2t}$ lie on parallel planes.

Differential equations of motion are as follows:

$$m_{1t}\ddot{x}_{2i-1} + kb_i(x_{2i-1} - x_{2i}) + kc_{i-1}(x_{2i-1} - x_{2i-2}) = 0 \tag{1.8.16a}$$

$$m_{2t}\ddot{x}_{2i} - kb_i(x_{2i-1} - x_{2i}) + kc_i(x_{2i} - x_{2i+1}) = 0;\ \ i = 1,2,\ldots,n \tag{1.8.16b}$$

For $n$ units in the rotationally periodic structure, $i+1 = 1$ when $i = n$, and $i-1 = n$ when $i = 1$. Define nondimensional time $t_n$ as

$$t_n = \omega_b t \tag{1.8.17a}$$

where

$$\omega_b = \sqrt{\frac{kb_t}{m_{1t}}} \tag{1.8.17b}$$

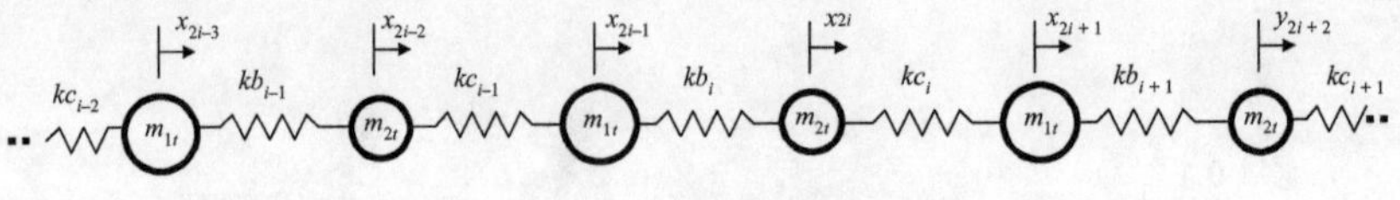

**Figure 1.8.8.**  One-dimensional nearly periodic structure with two atoms in each unit.

Also, define

$$kb_i = kb_t(1+r_{bi}); \quad kc_i = kc_t(1+r_{ci}) \qquad (1.8.18a, b)$$

where $kb_t$ and $kc_t$ are coupling stiffnesses of a perfectly tuned system, variables $r_{bi}$ and $r_{ci}$ represent disorder/mistuning in the system. Substituting Equations (1.8.17) and (1.8.18) into the Equation (1.8.16),

$$x''_{2i-1} + (1+r_{bi})(x_{2i-1} - x_{2i}) + \mu_k(1+r_{ci-1})(x_{2i-1} - x_{2i-2}) = 0 \qquad (1.8.19a)$$

$$\mu_m x''_{2i} - (1+r_{bi})(x_{2i-1} - x_{2i}) + \mu_k(1+r_{ci})(x_{2i} - x_{2i+1}) = 0 \qquad (1.8.19b)$$

where

$$x''_\ell = \frac{d^2 x_\ell}{dt_n^2}; \quad \mu_m = \frac{m_{2t}}{m_{1t}}; \quad \mu_k = \frac{kc_t}{kb_t} \qquad (1.8.20)$$

In matrix form, Equations (1.8.19a) and (1.8.19b) can be written as

$$M\mathbf{x}'' + K\mathbf{x} = 0 \qquad (1.8.21)$$

where

$$\mathbf{x} = \begin{bmatrix} \mathbf{z}_1 \\ \vdots \\ \mathbf{z}_i \\ \vdots \\ \mathbf{z}_n \end{bmatrix} \text{ and } \mathbf{z}_i = \begin{bmatrix} x_{2i-1} \\ x_{2i} \end{bmatrix} \qquad (1.8.22a, b)$$

$$M = \begin{bmatrix} M_u & & & \\ & M_u & & \\ & & \ddots & \\ & & & M_u \end{bmatrix}; \quad K = \begin{bmatrix} K_{u1} & K_{c1,2} & 0 & \cdots & K_{c1,n} \\ K_{c2,1} & K_{u2} & K_{c2,3} & \cdots & 0 \\ 0 & K_{c3,2} & K_{u3} & \cdots & 0 \\ \vdots & \vdots & \vdots & \vdots & \vdots \\ K_{cn,1} & 0 & 0 & \cdots & K_{un} \end{bmatrix} \qquad (1.8.23a, b)$$

$$M_u = \begin{bmatrix} 1 & 0 \\ 0 & \mu_m \end{bmatrix}; \quad K_{ui} = \begin{bmatrix} 1+r_{bi}+\mu_k(1+r_{ci-1}) & -(1+r_{bi}) \\ -(1+r_{bi}) & 1+r_{bi}+\mu_k(1+r_{ci}) \end{bmatrix} \qquad (1.8.24a, b)$$

$$K_{ci,i+1} = \begin{bmatrix} 0 & 0 \\ -\mu_k(1+r_{ci}) & 0 \end{bmatrix}; \quad K_{ci,i-1} = \begin{bmatrix} 0 & -\mu_k(1+r_{ci-1}) \\ 0 & 0 \end{bmatrix} \quad (1.8.25a, b)$$

## Perfectly Tuned System

For a perfectly tuned system, each set of coupling stiffnesses has identical values $kb_t$ or $kc_t$, that is,

$$r_{b1} = r_{b2} = \cdots = r_{bn} = 0 \tag{1.8.26}$$

$$r_{c1} = r_{c2} = \cdots = r_{cn} = 0 \tag{1.8.27}$$

With conditions (1.8.26) and (1.8.27), differential equations (1.8.19) can be written as

$$M_u \mathbf{z}_i'' + K_{ut}\mathbf{z}_i + K_{cR}\mathbf{z}_{i+1} + K_{cL}\mathbf{z}_{i-1} = \mathbf{0}; \quad i = 1, 2, ...., n \tag{1.8.28}$$

where

$$K_{ut} = \begin{bmatrix} 1+\mu_k & -1 \\ -1 & 1+\mu_k \end{bmatrix}; \quad K_{cR} = \begin{bmatrix} 0 & 0 \\ -\mu_k & 0 \end{bmatrix}; \quad K_{cL} = \begin{bmatrix} 0 & -\mu_k \\ 0 & 0 \end{bmatrix} \quad (1.8.29a, b, c)$$

$$\mathbf{z}_i = \begin{bmatrix} x_{2i-1} \\ x_{2i} \end{bmatrix} \tag{1.8.30}$$

Modal solutions of Equation (1.8.28) can be written to have the form of constant interdiatomic phase angles;

$$\mathbf{z}_i = \alpha e^{j(\omega_{tt\ell}t_n + \phi\ell(i-1))}; \quad \ell = 0, 1, 2, ...., n-1 \tag{1.8.31}$$

where $\alpha$ is a complex $2 \times 1$ vector, $\phi\ell$ is the constant interdiatomic phase angle, angle $\phi$ is defined by Equation (1.8.1), and $\omega_{tt\ell}$ is a nondimensional natural frequency defined as:

$$\omega_{tt\ell} = \frac{\omega_{t\ell}}{\omega_b} \tag{1.8.32}$$

Here, $\omega_{t\ell}$ is the actual tuned frequency, and $\omega_b$ is defined by Equation (1.8.17b). Substituting Equation (1.8.31) into Equation (1.8.28),

$$M_u \mathbf{z}_i'' + K_{eqt}\mathbf{z}_i = \mathbf{0} \tag{1.8.33}$$

where

$$K_{eqt} = K_{ut} + K_{cR}e^{j\phi\ell} + K_{cL}e^{-j\phi\ell} \tag{1.8.34}$$

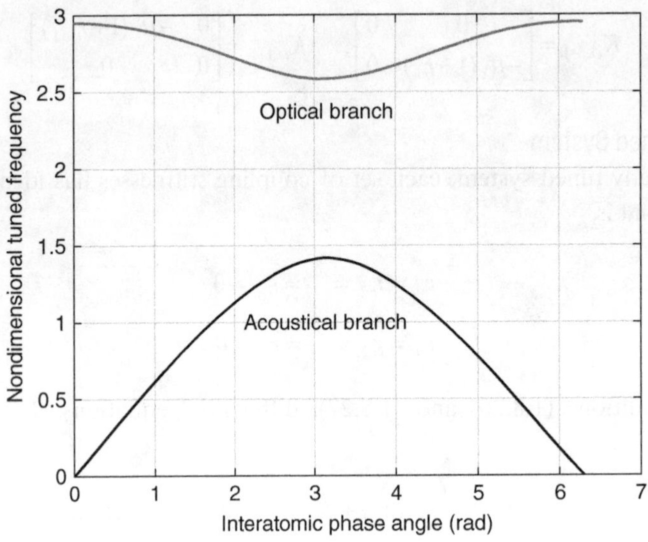



**Figure 1.8.9.**   Tuned natural frequencies for diatomic chain ($n = 300$, $\mu_m = 0.3$, $\mu_k = 1$).

Natural frequencies are found by setting $\det(K_{eqt} - \omega_{ttl}^2 M_u) = 0$. Result is

$$\omega_{ttl}^2 = \frac{(1+\mu_m)(1+\mu_k)\pm\sqrt{(1+\mu_m)^2(1+\mu_k)^2-16\mu_m\mu_k\sin^2(\phi\ell/2)}}{2\mu_m} \qquad (1.8.35)$$

Nondimensional frequencies are plotted as a function of interatomic phase angles $\phi\ell$; $\ell = 0,1,2,\ldots,n$, for a 300 atom ring, Figure 1.8.9, where mass ratio $\mu_m = 0.3$ and stiffness ratio $\mu_k = 1$. These frequencies are again plotted in the first Brillioun zone, Figure 1.8.10. There are two branches of frequencies. The lower and higher branches are known as acoustical and optical phonon branches (Kittel, 1996), respectively. The band of frequencies between the highest acoustical frequency and the lowest optical frequency is known as *bandgap* (Kittel, 1996).

### Disorder/Mistuning

Deviations (mistuning) in coupling stiffnesses $r_{bi}$ and $r_{ci}$ are chosen from a zero-mean normal distribution with standard deviation $= 0.07$. Natural frequencies (Figure 1.8.11) and corresponding modal vectors are obtained by assembling full-order mass and stiffness matrices and using the Matlab routine: *eig*. Again, all the repeated eigenvalues split due to mistuning, and there are unique eigenvectors. The participation ratio of each modal vector, Equation (1.8.13) modified for 2n dimensional vectors, is again computed and shown in Figure 1.8.12. The values of these participation ratios are smallest for modes near highest frequencies of both acoustical and optical branches, for example, Figure 1.8.13. Again, there are many more modes around these highest frequencies when compared to those around lower frequencies. In Figure 1.8.12, participation ratios are highest around mode number 100, implying that modes will not be localized, Figure 1.8.14.

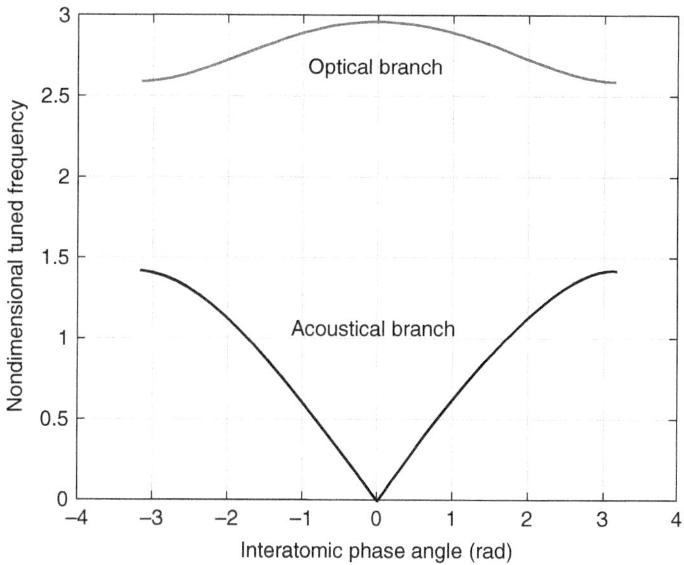



**Figure 1.8.10.**  Tuned natural frequencies for diatomic chain ($n = 300$, $\mu_m = 0.3$, $\mu_k = 1$) in a Brioullian zone.

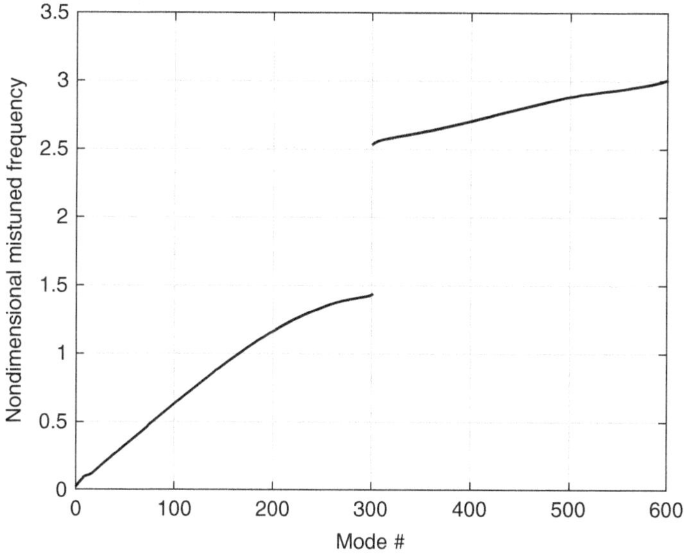



**Figure 1.8.11.**  Mistuned natural frequencies for diatomic chain ($n = 300$, $\sigma = 0.07$, $\mu_m = 0.3$, $\mu_k = 1$).

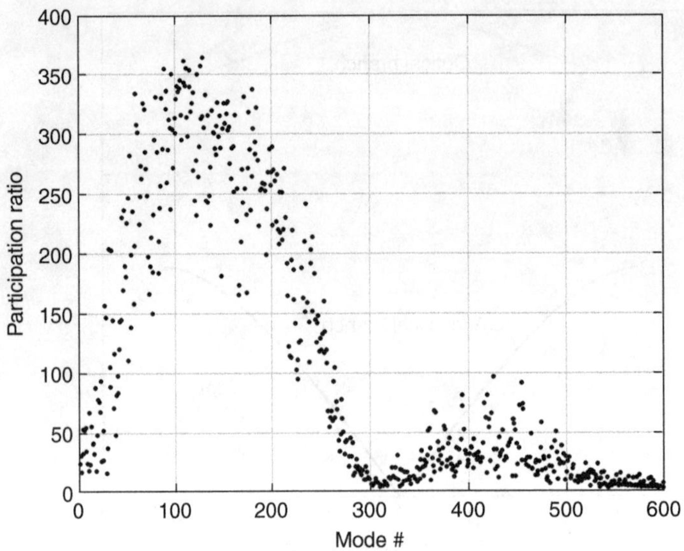

**Figure 1.8.12.** Participation ratio for mistuned diatomic chain ($n = 300$, $\sigma = 0.07$, $\mu_m = 0.3$, $\mu_k = 1$).

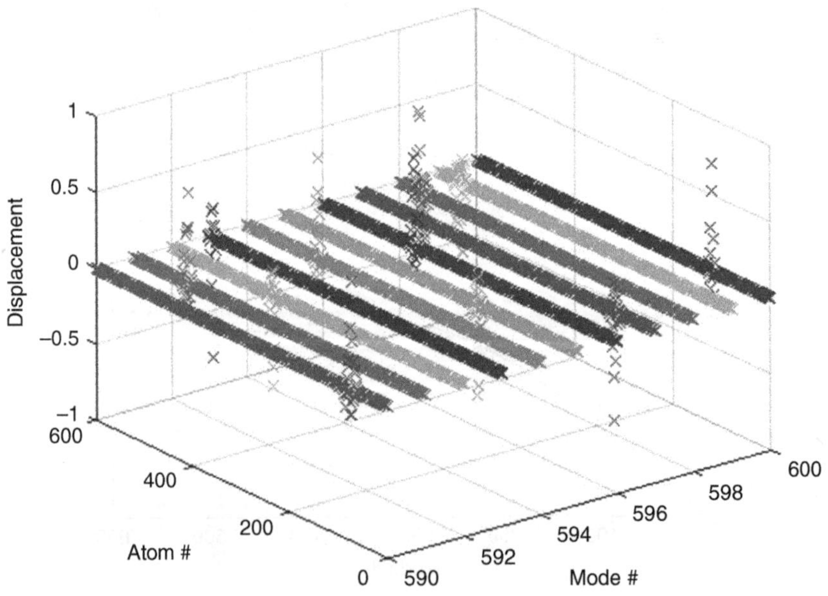

**Figure 1.8.13.** Localized modal vectors 595–600 of diatomic chain ($n = 300$, $\sigma = 0.07$, $\mu_m = 0.3$, $\mu_k = 1$).

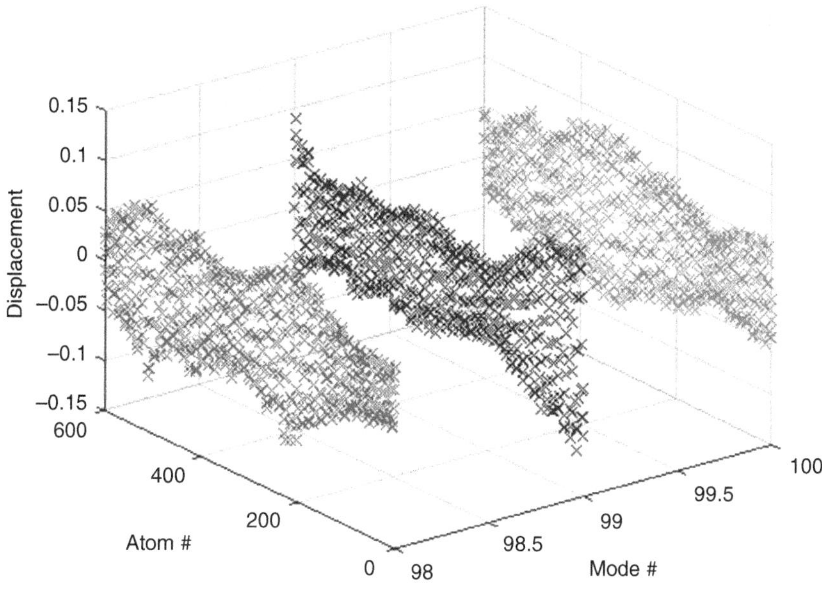

**Figure 1.8.14.** Nonlocalized modal vectors 98–100 of diatomic chain ($n = 300$, $\sigma = 0.07$, $\mu_m = 0.3$, $\mu_k = 1$).

# 2   Fundamentals of Forced Vibration of a Mistuned Rotationally Periodic Structure

First, reasons for steady-state amplitude amplification due to mistuning are explored. Modal approach is presented along with pole-zero cancellation phenomenon for a perfectly tuned system. Next, the following issues are discussed: computation of maximum amplitude amplification and its statistics, and the nature of probability distribution function of the maximum amplitude.

## 2.1   Basic Mistuned Model for Forced Vibration

The model shown in Figure 2.1.1 is obtained by adding a viscous damper and a force to each mass in Figure 1.1.1. Modal mass and stiffness of each blade are represented by $m_i$ and $k_i$, respectively. The structural coupling between adjacent blades due to the disk flexibility is represented by $k_c$.

The governing system of differential equations is represented by

$$m_i\ddot{x}_i + c_t\dot{x}_i + k_i x_i + k_c(x_i - x_{i+1}) + k_c(x_i - x_{i-1}) = f_i; \quad i = 1, 2, \ldots, n \quad (2.1.1)$$

In matrix form,

$$M\ddot{\mathbf{x}} + C_t\dot{\mathbf{x}} + K\mathbf{x}(t) = \mathbf{f}(t) \quad (2.1.2)$$

where

$$M = M_t + \Delta M \quad (2.1.3)$$

$$K = K_t + \Delta K \quad (2.1.4)$$

Matrices $M_t$, $K_t$, $\Delta M$, and $\Delta K$ are defined by Equations (1.1.7)–(1.1.10). The damping matrix $C_t$ and the forcing vector $\mathbf{f}(t)$ are defined as

$$C_t = c_t I_n \quad (2.1.5)$$

$$\mathbf{f}(t) = \begin{bmatrix} f_1(t) & f_2(t) & \cdots & f_n(t) \end{bmatrix}^T \quad (2.1.6)$$

An important scenario is that the (rotationally) periodic structure is rotating with a constant angular velocity $\omega$ in a nonuniform time-invariant pressure field. In this

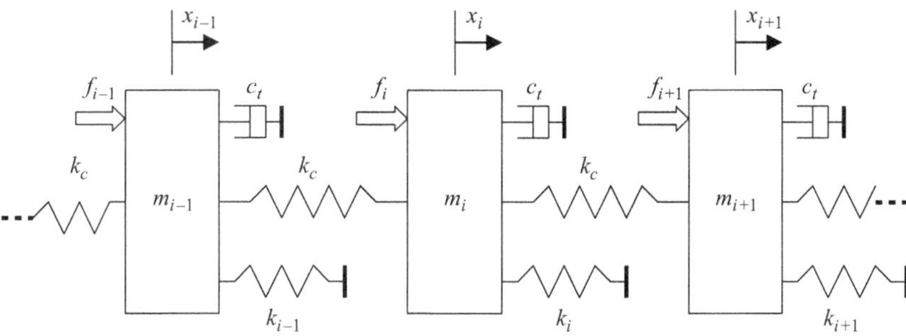


**Figure 2.1.1.** A mistuned rotationally periodic structure with damping and external excitation.

case, each mass will experience the same force, but with a phase difference. Further, this force will be periodic in nature, and can be expanded in a Fourier series.

Considering a single component in the Fourier series,

$$\mathbf{f}(t) = \mathbf{p}_\ell e^{j\omega t}; \; j = \sqrt{-1} \tag{2.1.7}$$

where the vector $\mathbf{p}_\ell$ is defined by Equation (1.2.5). The relationship (2.1.7) is an expression of the fact that each mass is subjected to the same magnitude of force, but with a constant blade-to-blade phase difference. In a bladed rotor, this type of excitation is called *engine order*.

## 2.1.1  Forced Response of a Tuned System including Pole-Zero Analysis

For a tuned system, $\Delta M = 0$ and $\Delta K = 0$. In steady state, the response is sinusoidal and can be expressed as a linear combination of eigenvectors:

$$\mathbf{x}(t) = \Phi \alpha_t e^{j\omega t} \tag{2.1.8}$$

where the matrix $\Phi$ is given by Equation (1.2.10), and the coefficients of eigenvectors are represented as

$$\alpha_t = \begin{bmatrix} \alpha_{t1} & \alpha_{t2} & \cdots & \alpha_{tn-1} & \alpha_{tn} \end{bmatrix}^T \tag{2.1.9}$$

Substituting Equations (2.1.7) and (2.1.8) into Equation (2.1.2),

$$(-\omega^2 m_t I + j\omega c_t I + K_t)\Phi \alpha_t = \mathbf{p}_\ell \tag{2.1.10}$$

Premultiplying both sides of Equation (2.1.10) by the complex conjugate transpose of $\Phi$, $\Phi^H$,

$$\Phi^H(-\omega^2 m_t I + j\omega c_t I + K_t)\Phi \alpha_t = \Phi^H \mathbf{p}_\ell \tag{2.1.11}$$

Using properties (1.2.7) and (1.2.8),

$$(-\omega^2 I + \Omega^2 + j\omega c_t m_t^{-1} I)\alpha_t = \Phi^H \mathbf{p}_\ell \tag{2.1.12}$$

Using definitions (1.2.9) and (1.2.10),

$$(-\omega^2 + \omega_{fi}^2 + j\omega c_t m_t^{-1})\alpha_{ti} = \mathbf{p}_i^H \mathbf{p}_\ell \tag{2.1.13}$$

Because the matrix $\Phi$ is orthonormal,

$$\alpha_{t\ell} = \frac{1}{(-\omega^2 + \omega_{f\ell}^2 + j\omega c_t m_t^{-1})} \tag{2.1.14}$$

and

$$\alpha_{ti} = 0; \quad i \neq \ell \tag{2.1.15}$$

Therefore, the resonance occurs only when $\omega = \omega_{f\ell}$.

Next, the input/output description of the blade response under engine-order excitation is described. Let the forcing function be described as

$$\mathbf{f}(t) = \mathbf{p}_\ell g(t) \tag{2.1.16}$$

where $g(t)$ is a scalar sinusoidal forcing function, which will serve as the input. Let the output be the displacement of the mass number $i$,

$$y_i(t) = x_i(t) = \mathbf{c}_{oi}\mathbf{x}(t) \tag{2.1.17}$$

where only the element number $i$ of the vector $\mathbf{c}_{oi}$ is nonzero, that is,

$$\mathbf{c}_{oi} = \begin{bmatrix} 0 & \cdots & 0 & 1 & 0 & \cdots & 0 \end{bmatrix} \tag{2.1.18}$$

Taking Laplace transforms of Equations (2.1.2) and (2.1.17) with zero initial conditions, the transfer function (Sinha, 2007) is defined as follows:

$$\frac{y_i(s)}{g(s)} = \mathbf{c}_{oi}(s^2 I + m_t^{-1}c_t sI + m_t^{-1}K_t)^{-1} m_t \mathbf{p}_\ell \tag{2.1.19}$$

or

$$\frac{y_i(s)}{g(s)} = \frac{\mathbf{c}_{oi}\, adj(s^2 I + m_t^{-1}c_t sI + m_t^{-1}K_t) m_t \mathbf{p}_\ell}{\det(s^2 I + m_t^{-1}c_t sI + m_t^{-1}K_t)} = \frac{\eta_i(s)}{\lambda(s)} \tag{2.1.20}$$

where the symbol *adj* denotes adjugate or adjoint of a square matrix. The zeros of the transfer function (Jones and Cross, 2003) are the roots of the numerator polynomial $\eta_i(s) = 0$, whereas the poles of the transfer function are the roots of the denominator polynomial $\lambda(s) = 0$. For a perfectly tuned system, displacements of all blades are identical. Therefore,

$$\eta_1(s) = \eta_2(s) = \cdots = \eta_n(s) \tag{2.1.21}$$

**Table 2.1.1.** Poles and zeros (Sinha, 2008a) of the tuned system (scaled by $\sqrt{(k_t + 4k_c)/m_t}$) ($k_t = 430000$, $k_c = 4543$, $c = 0.138$, $\ell = 1$)

| Poles | Zeros |
|---|---|
| $-0.0010 + 0.9795i$ | $-0.0010 + 0.9795i$ |
| $-0.0010 - 0.9795i$ | $-0.0010 - 0.9795i$ |
| $-0.0010 + 0.9815i$ | $-0.0010 + 0.9815i$ |
| $-0.0010 - 0.9815i$ | $-0.0010 - 0.9815i$ |
| $\mathbf{-0.0010 + 0.9815i}$ | |
| $\mathbf{-0.0010 - 0.9815i}$ | |
| $-0.0010 + 0.9866i$ | $-0.0010 + 0.9866i$ |
| $-0.0010 - 0.9866i$ | $-0.0010 - 0.9866i$ |
| $-0.0010 + 0.9866i$ | $-0.0010 + 0.9866i$ |
| $-0.0010 - 0.9866i$ | $-0.0010 - 0.9866i$ |
| $-0.0010 + 0.9930i$ | $-0.0010 + 0.9930i$ |
| $-0.0010 - 0.9930i$ | $-0.0010 - 0.9930i$ |
| $-0.0010 + 0.9930i$ | $-0.0010 + 0.9930i$ |
| $-0.0010 - 0.9930i$ | $-0.0010 - 0.9930i$ |
| $-0.0010 + 0.9981i$ | $-0.0010 + 0.9981i$ |
| $-0.0010 - 0.9981i$ | $-0.0010 - 0.9981i$ |
| $-0.0010 + 0.9981i$ | $-0.0010 + 0.9981i$ |
| $-0.0010 - 0.9981i$ | $-0.0010 - 0.9981i$ |
| $-0.0010 + 1.0000i$ | $-0.0010 + 1.0000i$ |
| $-0.0010 - 1.0000i$ | $-0.0010 - 1.0000i$ |

From the Equations (2.1.14) and (2.1.15), it is clear that the effective order of the system is two. Because the order of the denominator polynomial equation $\lambda(s) = 0$ is $2n$, there are $2(n-1)$ pole-zero cancellations; for example, Table 2.1.1 for $n = 10$. In control theory, pole-zero cancellations are related to loss of observability (Sinha, 2007), which is indeed happening here because the output contains only one mode of vibration.

For the frequency response, the excitation force is assumed to be sinusoidal, that is,

$$g(t) = g_0 e^{\iota \omega t}; \qquad \iota = \sqrt{-1} \tag{2.1.22}$$

where $g_0$ is a real number. Then the steady-state response can be written as

$$y_i(t) = y_{c0i} e^{\iota \omega t} \tag{2.1.23}$$

where $y_{c0i}$ is a complex number and

$$\frac{|y_{c0i}|}{g_0} = \frac{|\eta_i(\iota \omega)|}{|\lambda(\iota \omega)|} = |h_i(\iota \omega)|; \qquad \iota = \sqrt{-1} \tag{2.1.24}$$

## 2.2          Forced Response of a Mistuned System

### 2.2.1          Transfer of Energy to Many Modes and Pole/Zero Analysis

Consider the engine-order excitation (2.1.7) again, that is,

$$\mathbf{f}(t) = \mathbf{p}_\ell e^{j\omega t} \tag{2.2.1}$$

In steady state, the response is sinusoidal and can be again expressed as a linear combination of tuned eigenvectors:

$$\mathbf{x}(t) = \Phi \alpha_t e^{j\omega t} \tag{2.2.2}$$

where the matrix $\Phi$ is given by Equation (1.2.10), and $\alpha_t$ by Equation (2.1.9). Substituting Equations (2.2.1) and (2.2.2) into Equation (2.1.2),

$$(-\omega^2 m_t I + \Delta M + j\omega c_t I + K_t + \Delta K)\Phi \alpha_t = \mathbf{p}_\ell \tag{2.2.3}$$

Premultiplying both sides of Equation (2.2.3) by $\Phi^H$,

$$\Phi^H(-\omega^2 m_t I - \omega^2 \Delta M + j\omega c_t I + K_t + \Delta K)\Phi \alpha_t = \Phi^H \mathbf{p}_\ell \tag{2.2.4}$$

Using properties (1.2.7) and (1.2.8),

$$(-\omega^2 I + \Omega^2 + j\omega c_t m_t^{-1} I)\alpha_t - \omega^2 \Phi^H \Delta M \Phi \alpha_t + \Phi^H \Delta K \Phi \alpha_t = \Phi^H \mathbf{p}_\ell \tag{2.2.5}$$

Because the matrix $\Phi$ is orthonormal,

$$\Phi^H \mathbf{p}_\ell = \begin{bmatrix} \mathbf{p}_0^H \\ \mathbf{p}_1^H \\ \mathbf{p}_2^H \\ \vdots \\ \mathbf{p}_{n-1}^H \end{bmatrix} \mathbf{p}_\ell = \begin{bmatrix} 0 \\ \vdots \\ 0 \\ \mathbf{p}_\ell^H \mathbf{p}_\ell \\ 0 \\ \vdots \\ 0 \end{bmatrix} = \begin{bmatrix} 0 \\ \vdots \\ 0 \\ 1 \\ 0 \\ \vdots \\ 0 \end{bmatrix} \tag{2.2.6}$$

Let

$$-\omega^2 \Phi^H \Delta M \Phi = A \quad \text{and} \quad \Phi^H \Delta K \Phi = B \tag{2.2.7a, b}$$

Using definitions (1.2.9) and (1.2.10),

$$(-\omega^2 + \omega_{f\ell}^2 + j\omega c_t m_t^{-1})\alpha_{t\ell} + \sum_{i=1}^{n}(a_{\ell i} + b_{\ell i})\alpha_{ti} = 1 \tag{2.2.8a}$$

$$(-\omega^2 + \omega_{fp}^2 + j\omega c_t m_t^{-1})\alpha_{tp} + \sum_{i=1}^{n}(a_{pi} + b_{pi})\alpha_{ti} = 0; \quad p = 1, 2, \cdots, \ell-1, \ell+1, \ldots, n \tag{2.2.8b}$$

**Table 2.2.1.** Poles and zeros (Sinha, 2008a) of a mistuned system (scaled by $\sqrt{(k_t + 4k_c)/m_t}$) ($k_c = 4543$, $c = 0.138$, $\ell = 1$), mistuned disk no. 1,000

| Poles | Zeros |
|---|---|
| $-0.0010 + 0.9767i$ | $-0.0007 + 0.9773i$ |
| $-0.0010 - 0.9767i$ | $-0.0012 - 0.9773i$ |
| $-0.0010 + 0.9780i$ | $-0.0022 + 0.9792i$ |
| $-0.0010 - 0.9780i$ | $0.0003 - 0.9792i$ |
| $-0.0010 + 0.9810i$ | |
| $-0.0010 - 0.9810i$ | |
| $-0.0010 + 0.9843i$ | $-0.0009 + 0.9844i$ |
| $-0.0010 - 0.9843i$ | $-0.0010 - 0.9844i$ |
| $-0.0010 + 0.9861i$ | $-0.0022 + 0.9872i$ |
| $-0.0010 - 0.9861i$ | $0.0002 - 0.9872i$ |
| $-0.0010 + 0.9910i$ | $0.0021 + 0.9902i$ |
| $-0.0010 - 0.9910i$ | $-0.0040 - 0.9902i$ |
| $-0.0010 + 0.9921i$ | $-0.0011 + 0.9914i$ |
| $-0.0010 - 0.9921i$ | $-0.0009 - 0.9914i$ |
| $-0.0010 + 0.9961i$ | $-0.0009 + 0.9960i$ |
| $-0.0010 - 0.9961i$ | $-0.0010 - 0.9960i$ |
| $-0.0010 + 0.9969i$ | $-0.0019 + 0.9969i$ |
| $-0.0010 - 0.9969i$ | $-0.0000 - 0.9969i$ |
| $-0.0010 + 1.0003i$ | $-0.0010 + 1.0003i$ |
| $-0.0010 - 1.0003i$ | $-0.0009 - 1.0003i$ |

where $a_{pi}$ and $b_{pi}$ are elements of matrices $A$ and $B$, respectively. Compared to the tuned system response, all the coefficients of tuned modes $\alpha_{ti}$ are nonzero, and will be obtained after solving Equations (2.2.8a) and (2.2.8b) simultaneously. In other words, even though the excitation corresponds to the tuned mode number $\ell$, energy gets transferred to all tuned modes.

Again, the input/output description of the blade response under engine-order excitation is described. Let the forcing function be described by Equation (2.1.16) and the output $y_i$ by Equations (2.1.17) and (2.1.18). Taking Laplace transforms of Equations (2.1.2), (2.1.16), and (2.1.17) with zero initial conditions, the transfer function (Sinha, 2007) is defined as follows:

$$\frac{y_i(s)}{g(s)} = \mathbf{c}_{oi}(s^2 I + M^{-1}c_t sI + M^{-1}K)^{-1} M^{-1}\mathbf{p}_\ell \tag{2.2.9}$$

or

$$\frac{y_i(s)}{g(s)} = \frac{\mathbf{c}_{oi}\,adj(s^2 I + M^{-1}c_t sI + M^{-1}K)^{-1} M^{-1}\mathbf{p}_\ell}{\det(s^2 I + M^{-1}c_t sI + M^{-1}K)} = \frac{\eta_i(s)}{\lambda(s)} \tag{2.2.10}$$

In the case of a mistuned system, the numerator polynomial is different for each blade, and the variations in the blade amplitudes are related to the variations in the locations of zeros. Further, there is no pole-zero cancellation (see Table 2.2.1). It should also be noted that the vector $\mathbf{p}_\ell$ contains complex elements for all interblade phase angles except 0 and 180 degrees. Therefore, complex zeros need not have their conjugates as zeros, that is, locations of zeros will not be symmetric with respect to the real axis.

The state-space model (Sinha, 2007) for the system (2.1.2) with input (2.1.16) and output (2.1.17) can be described as

$$\dot{\mathbf{z}} = A\mathbf{z} + \mathbf{b}g(t) \tag{2.2.11a}$$

$$y_i(t) = \mathbf{c}_0\mathbf{z}(t) \tag{2.2.11b}$$

where

$$\mathbf{z}(t) = \begin{bmatrix} \mathbf{x}(t) \\ \dot{\mathbf{x}}(t) \end{bmatrix} \tag{2.2.12}$$

$$A = \begin{bmatrix} \mathbf{0} & I_n \\ -M^{-1}K & -M^{-1}C_t \end{bmatrix} \tag{2.2.13}$$

$$\mathbf{b} = \begin{bmatrix} \mathbf{0}_{n\times 1} \\ M^{-1}\mathbf{p}_\ell \end{bmatrix} \tag{2.2.14}$$

and

$$\mathbf{c}_o = \begin{bmatrix} \mathbf{c}_{oi} & \mathbf{0}_{1\times n} \end{bmatrix} \tag{2.2.15}$$

For the frequency response, the excitation force is assumed to be sinusoidal, that is,

$$g(t) = g_0 \sin \omega t \tag{2.2.16}$$

Then the steady-state response can be written as

$$y_i(t) = y_{0i} \sin(\omega t + \theta_i) \tag{2.2.17}$$

where $\theta_i$ is the phase difference between output and input, and

$$\frac{y_{0i}}{g_0} = \frac{|\eta_i(j\omega)|}{|\lambda(j\omega)|} = |h_i(j\omega)| \tag{2.2.18}$$

There are four types of amplitude amplifications for mistuning analysis:

$$I_1 = y_{0r1} \text{ at a fixed excitation frequency } \omega \tag{2.2.19}$$

$$I_2 = \max_i y_{0ri} \text{ at a fixed excitation frequency } \omega \tag{2.2.20}$$

$$I_3 = \max_{\omega} \; y_{0r1} \tag{2.2.21}$$

and

$$I_4 = \max_{\omega} \max_{i} \; y_{0ri} \tag{2.2.22}$$

where $y_{0ri}$ is the ratio of $y_{0i}$ and the resonant tuned system amplitude.

Variations in modal parameters of blades are random variables (Sinha, 1986). For any defined mean and standard deviations of these random variables, there exist theoretically an infinite number of possible combinations of these random variables. Because of the cyclic nature of a bladed disk, the maximum values of $I_1$ and $I_2$ will be theoretically identical, that is, for a large number of simulations. Similarly, the maximum values of $I_3$ and $I_4$ will also be identical. However, minimum values and statistical distributions of $I_1$ and $I_2$, and those of $I_3$ and $I_4$ will not be identical. The objective function $I_3$ is the $H_\infty$ norm of the single-input/single-output (SISO) transfer function $h_i(s)$, and can be computed using the MATLAB routine, which is based on the bounded real lemma (Sinha, 2007):

$$\left\| h_i(s) \right\|_\infty < \gamma \tag{2.2.23}$$

if and only if the Hamiltonian matrix

$$H = \begin{bmatrix} A & \dfrac{1}{\gamma^2}\,\mathbf{b}\mathbf{b}^T \\ -\mathbf{c}_o^T\mathbf{c}_o & -A^T \end{bmatrix} \tag{2.2.24}$$

has no eigenvalues on the imaginary axis. The matrices $A$, $\mathbf{b}$, and $\mathbf{c}_o$ in Equation (2.2.24) describe the state-space model, Equation (2.2.11).

This bounded real lemma is a quick way to check if the amplitude exceeds a critical value $\gamma$ without evaluating frequency responses at many points on the frequency grid. Furthermore, if the frequency grid is not fine enough for a small amount of damping, which is typically the case, the predicted maximum value of the frequency response can deviate significantly from the actual value. This is where the bounded real lemma can be used to avoid a large error in the computation of the maximum amplitude.

### Connection between Zeros and Mistuned Modal Vectors

Let $\boldsymbol{\psi}_i$ and $\omega_i$ be the modal vector and the natural frequency from the solution of the eigenvalue/eigenvector problem of the mistuned system, that is,

$$K\boldsymbol{\psi}_i = \omega_i^2 M \boldsymbol{\psi}_i \tag{2.2.25}$$

From Equation (2.2.25),

$$\Psi^T K \Psi = \Psi \Lambda \tag{2.2.26}$$

because

$$\Psi^T M \Psi = I_n \tag{2.2.27}$$

where

$$\Lambda = diag.[\omega_1^2 \quad \omega_2^2 \quad \cdots \quad \omega_n^2] \tag{2.2.28}$$

and

$$\Psi = [\psi_1 \quad \psi_2 \quad \cdots \quad \psi_n] \tag{2.2.29}$$

Expressing the displacement vector $\mathbf{x}(t)$ as a linear combination of mistuned modal vectors $\psi_i; i = 1, 2, \ldots, n,$

$$\mathbf{x}(t) = \Psi \alpha(t) \tag{2.2.30}$$

where the elements of the $n \times 1$ vector $\alpha$ are the coefficients of the mistuned modal vectors. Let

$$\alpha(t) = [\alpha_1(t) \quad \alpha_2(t) \quad \cdots \quad \alpha_n(t)]^T \tag{2.2.31}$$

Substituting Equations (2.2.30) and (2.1.16) into Equation (2.1.2), and using Equations (2.2.26)–(2.2.29),

$$\ddot{\alpha}_j + 2\xi_j \omega_j \dot{\alpha}_j + \omega_j^2 \alpha_j = q_j g(t); \quad j = 1, 2, \ldots, n \tag{2.2.32}$$

where $q_j$ is the *jth* element of the $n \times 1$ vector $\mathbf{q}$ defined as follows:

$$\mathbf{q} = \Psi^T \mathbf{p}_\ell = [q_1 \quad q_2 \quad \cdots \quad q_n]^T \tag{2.2.33}$$

and $\xi_j$ is the modal damping ratio. Taking Laplace transform of the Equation (2.2.32) with zero initial conditions:

$$\alpha_j(s) = \frac{q_j g(s)}{s^2 + 2\xi_j \omega_j s + \omega_j^2} \tag{2.2.34}$$

Using Equations (2.1.17) and (2.2.30),

$$y_i(t) = \beta_i \alpha(t) \tag{2.2.35}$$

where the $1 \times n$ vector $\beta_i$ is defined as follows:

$$\beta_i = \mathbf{c}_{oi} \Psi = [\beta_{i1} \quad \beta_{i2} \quad \cdots \quad \beta_{in}] \tag{2.2.36}$$

Equation (2.2.35) can also be written as

$$y_i(t) = \sum_{j=1}^{n} \beta_{ij} \alpha_j(t) \tag{2.2.37}$$

From Equations (2.2.32) and (2.2.37),

$$\frac{y_i(s)}{g(s)} = \sum_{j=1}^{n} \frac{\beta_{ij} q_j}{(s^2 + 2\xi_j \omega_j s + \omega_j^2)} \tag{2.2.38}$$

With the sinusoidal $g(t)$, Equation (2.1.22), the complex amplitude of the steady-state output, Equation (2.1.23) is expressed as a function of the excitation frequency $\omega$ and is obtained by substituting $s = \iota\omega$ on the right-hand side of Equation (2.2.38):

$$\frac{y_{c0i}}{g_0} = \sum_{j=1}^{n} \frac{\beta_{ij} q_j}{(\omega_j^2 - \omega^2) + \iota 2\xi_j \omega_j \omega}; \quad \iota = \sqrt{-1} \tag{2.2.39}$$

Multiplying numerators and denominators of terms inside the summation sign by the complex conjugate of the denominator,

$$\frac{y_{0i}}{g_0} = u_{xi} + \iota u_{yi} \tag{2.2.40}$$

where

$$u_{xi} = \sum_{j=1}^{n} \frac{u_{ijR}}{v_j} \tag{2.2.41}$$

$$u_{yi} = \sum_{j=1}^{n} \frac{u_{ijI}}{v_j} \tag{2.2.42}$$

$$v_j = [(\omega_j^2 - \omega^2)^2 + (2\xi_j \omega_j \omega)^2] \tag{2.2.43}$$

$$u_{ijR} = \beta_{ij}(q_{jR}(\omega_j^2 - \omega^2) + q_{jI} 2\xi_j \omega_j \omega) \tag{2.2.44}$$

$$u_{ijI} = \beta_{ij}(q_{jI}(\omega_j^2 - \omega^2) - q_{jR} 2\xi_j \omega_j \omega) \tag{2.2.45}$$

Note that $q_{jR}$ and $q_{jI}$ are real and imaginary parts of $q_j$, that is,

$$q_j = q_{jR} + \iota q_{jI} \tag{2.2.46}$$

### Case I ($0^0$ and $180^0$ Tuned Modes)

Let the constant interblade phase angle of the forcing function be either $0^0$ or $180^0$. In this case, the matrix $\mathbf{p}_\ell$ in Equation (2.1.16) will be composed of real numbers only. Furthermore, assume that the mistuned mode shape corresponding to the perturbations in tuned modes ($0^0$ or $180^0$) be $\mathbf{\psi}_p$. Then, on the basis of the approximation of Equation (2.2.30) as $\mathbf{x} \cong \mathbf{\psi}_p \alpha_p$, the Equation (2.2.40) is approximated as

$$\frac{y_{c0i}}{g_0} = \frac{u_{ipR} + \iota u_{ipI}}{v_p} \tag{2.2.47}$$

Substituting Equations (2.2.43)–(2.2.45) into Equation (2.2.47),

$$\frac{\left|y_{c0i}\right|}{g_0} = \frac{\left|\beta_{ip}q_{pR}\right|}{\sqrt{[(\omega_p^2 - \omega^2)^2 + (2\xi_p\omega_p\omega)^2]}} \tag{2.2.48}$$

### Case II (Repeated Tuned Modes)

Let the constant interblade phase angle of the forcing function be neither $0^0$ nor $180^0$. In this case, the matrix $\mathbf{p}_\ell$ in Equation (2.1.16) will be composed of complex numbers. Furthermore, assume that the mistuned mode shapes corresponding to the perturbations in both tuned modes be $\mathbf{\psi}_\sigma$ and $\mathbf{\psi}_{\sigma+1}$. Then, on the basis of the approximation of Equation (2.2.30) as $\mathbf{x} \cong \mathbf{\psi}_\sigma\alpha_\sigma + \mathbf{\psi}_{\sigma+1}\alpha_{\sigma+1}$, the Equation (2.2.40) is approximated as

$$\frac{y_{c0i}}{g_0} = \sum_{j=\sigma}^{\sigma+1} \frac{u_{ijR} + iu_{ijI}}{[(\omega_j^2 - \omega^2)^2 + (2\xi_j\omega_j\omega)^2]} \tag{2.2.49}$$

Substituting Equations (2.2.44) and (2.2.45) into Equation (2.2.49),

$$\frac{\left|y_{c0i}\right|^2}{g_0^2} = \sum_{j=\sigma}^{\sigma+1} \frac{\beta_{ij}^2(q_{jR}^2 + q_{jI}^2)}{[(\omega_j^2 - \omega^2)^2 + (2\xi_j\omega_j\omega)^2]} + \frac{2\beta_{ij}\prod_{j=\sigma}^{\sigma+1}(q_{jR}(\omega_j^2 - \omega^2) + q_{jI}2\xi_j\omega_j\omega)}{\prod_{j=\sigma}^{\sigma+1}[(\omega_j^2 - \omega^2)^2 + (2\xi_j\omega_j\omega)^2]}$$

$$+ \frac{2\beta_{ij}\prod_{j=\sigma}^{\sigma+1}(q_{jI}(\omega_j^2 - \omega^2) - q_{jR}2\xi_j\omega_j\omega)}{\prod_{j=\sigma}^{\sigma+1}[(\omega_j^2 - \omega^2)^2 + (2\xi_j\omega_j\omega)^2]} \tag{2.2.50}$$

### Perturbation Analysis

From Equation (2.2.49),

$$\chi_i = \frac{u_i}{v_i} \tag{2.2.51}$$

where $\chi_i = \left|y_{c0i}\right|/g_0$. Let the joint distribution of $u_i$ and $v_i$ be $f_{uv}(u_i,v_i)$. Then, the probability density function of $\chi_i$ can be shown (Papoulis, 1965) to be

$$f_\chi(\chi_i) = \int_0^\infty v_i f_{uv}(\chi_i v_i, v_i)dv_i \tag{2.2.52}$$

In other words, if the joint distribution $f_{uv}(u_i,v_i)$ is known, the probability density function of the amplitude, $\chi_i$, can be computed using a simple one-dimensional integration.

Expanding $u_i$ and $v_i$ in a Taylor series and neglecting terms higher than second-order,

$$u_i \cong u_t + \sum_{j=1}^{n}\left(\frac{\partial u_i}{\partial k_j}\right)\delta k_j + \frac{1}{2}\sum_{\ell=1}^{n}\sum_{j=1}^{n}\left(\frac{\partial^2 u_i}{\partial k_\ell \partial k_j}\right)\delta k_\ell \delta k_j \qquad (2.2.53)$$

$$v_i \cong v_t + \sum_{j=1}^{n}\left(\frac{\partial v_i}{\partial k_j}\right)\delta k_j + \frac{1}{2}\sum_{\ell=1}^{n}\sum_{j=1}^{n}\left(\frac{\partial^2 v_i}{\partial k_\ell \partial k_j}\right)\delta k_\ell \delta k_j \qquad (2.2.54)$$

Sinha and Chen (1989) have shown that the most of the random variables $\delta k_j$ and $\delta k_\ell \delta k_j$ are uncorrelated. Therefore, on the basis of the central limit theorem, it is possible that $u_i$ and $v_i$ are jointly Gaussian, which depends on $E(u_i)$, $E(v_i)$, $E(u_i^2)$, $E(v_i^2)$, and $E(u_i v_i)$.

## 2.2.2 Numerical Results Including Monte Carlo Statistics

First, statistical distributions or histograms of $I_1, I_2, I_3,$ and $I_4$, Equations (2.2.19)–(2.2.22), obtained from Monte Carlo simulation are presented in Figure 2.2.1. The number of blades, $n$, is chosen to be 10, $m_t = 0.0114$ kg., $k_t = 430000$ N/m, $k_c = 4543$ N/m, and $c = 0.138$ N-sec./m(damping ratio = 0.1 percent). The interblade phase angle of excitation is $108^0$, that is, $\ell = 3$ for the interblade phase angle $\phi\ell$ in Equation (1.2.5). The number of simulation ($ns$) is 10,000, with the standard deviation ($\sigma_k$) of $k_i$ equal to 10,000 N/m, which represents about 1 percent mistuning in terms of ratio of standard deviation and mean of blades' natural frequencies. For each simulation, $I_3$ and $I_4$ are obtained by evaluating amplitudes at $nog = 1,000$ frequencies equispaced between $\omega_\ell = 0.95\omega_r$ and $\omega_u = 1.05\omega_r$, where $\omega_r$ is the resonant frequency of the tuned system. The values of $I_1$ and $I_2$ are obtained for the excitation frequency $\omega$ equal to $\omega_r$. The maximum values of $I_1$ and $I_2$ are close to each other, but minimum values and statistical distributions are quite different. Similar results are found for $I_3$ and $I_4$. Hence, it is important to use $I_2$ and $I_4$, which involves computation of the infinity norm of the amplitude vector, to find the minimum value and statistical distribution of the amplitude amplification due to mistuning.

Next, the validity of Equation (2.2.48) is examined (see Figures 2.2.2 and 2.2.3). The number of simulations is 1,000, with the standard deviation of $k_i$ equal to 2,000 N/m. For each simulation, $I_4$ is obtained by evaluating amplitudes at 10,000 frequencies equispaced between $0.95\omega_r$ and $1.05\omega_r$, where $\omega_r$ is the resonant frequency of the tuned system. Amplitude ratio is defined as the ratio of peak maximum amplitude ($I_4$) predicted on the basis of Equation (2.2.48), and $I_4$ predicted on the basis of all modes of vibration. Frequency ratio is defined as the ratio of the excitation frequency for $I_4$ and the resonant natural frequency of the tuned system. Results in Figure 2.2.2 indicate that the amplitude ratio is around one, and $I_4$ can be only predicted based on a perturbed 180 degree tuned mode. Figure 2.2.3 also

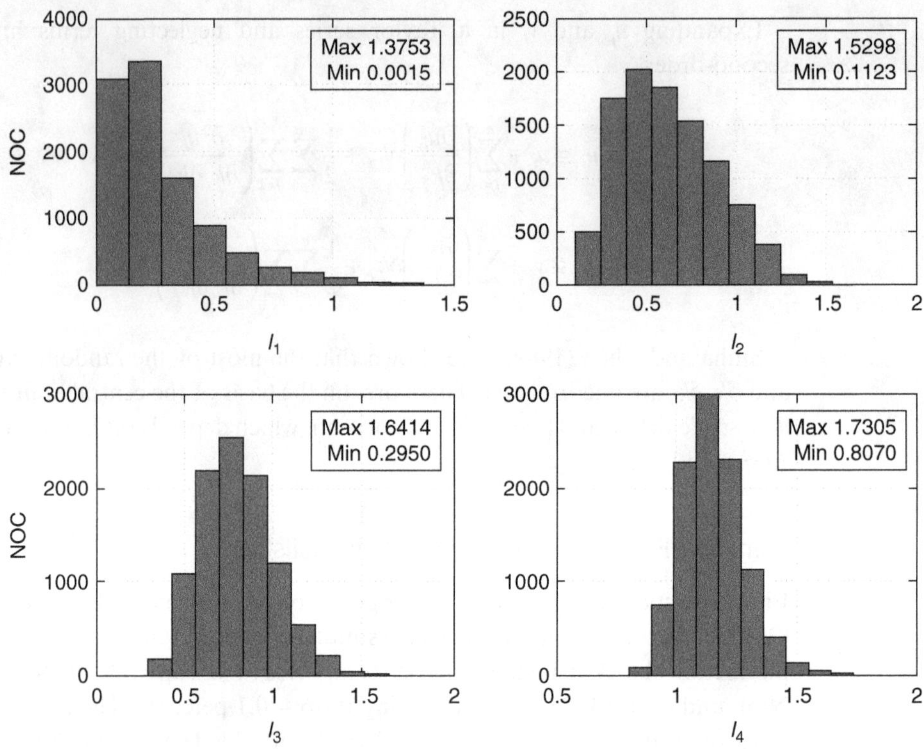

**Figure 2.2.1.**   Statistical distributions of four amplitude amplification measures ($I_1$, $I_2$, $I_3$, and $I_4$) ($k_c = 4543$, $c = 0.138$, $ns = 10000$, $nog = 1000$, $\omega_\ell = 0.95\omega_r$, $\omega_u = 1.05\omega_r$, $\ell = 3$ (108 degrees), no. of blades = 10, $\sigma_k = 10000$) $\omega_r$ is the natural frequency of tuned system (Sinha, 2008a).

indicates the frequency at which the peak maximum amplitude occurs is almost equal to the natural frequency of the mistuned bladed disk.

Next, a case of repeated modes is considered. The value of $\ell$ is chosen to be 3, which corresponds to $108^0$ interblade phase angle for excitation, and a pair of modes (number 6 and number 7). The Equation (2.2.50) is quite valid (see Figure 2.2.4). However, excitation frequencies for the peak maximum amplitudes can be almost equal to higher mistuned natural frequency, or to lower mistuned frequency, or be somewhere between the lower and higher natural frequencies (see Figure 2.2.5). The contribution of each mode (number 6 and number 7) also varies a great deal (see Figures 2.2.6 and 2.2.7), which clearly suggests that both modes must be considered in estimating the response.

Next, statistical distributions of numerators and denominators of the transfer functions corresponding to peak maximum amplitudes are examined (see Figure 2.2.8). Magnitudes of numerators and denominators have been scaled with respect to their tuned values. It is interesting to note that the magnitude of the denominator can change by a factor of about 18, whereas the maximum change in the numerator is about twenty-eight-fold. This suggests that changes in zeros are larger than those in poles. Because zeros are related to mistuned

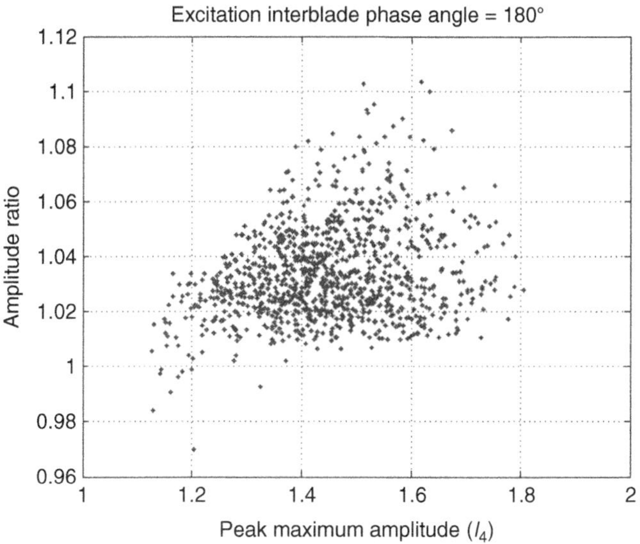

**Figure 2.2.2.**  Contribution of the resonant mode to the peak maximum amplitude ($k_c = 4543$, $c = 0.138$, $ns = 1000$, $nog = 10000$, $\omega_\ell = 0.95\omega_r$, $\omega_u = 1.05\omega_r$, $\ell = 5$ (180 degrees), no. of blades = 10, $\sigma_k = 2000$), frequency grid is fixed and based on $\omega_r$ of tuned system (Sinha, 2008a).

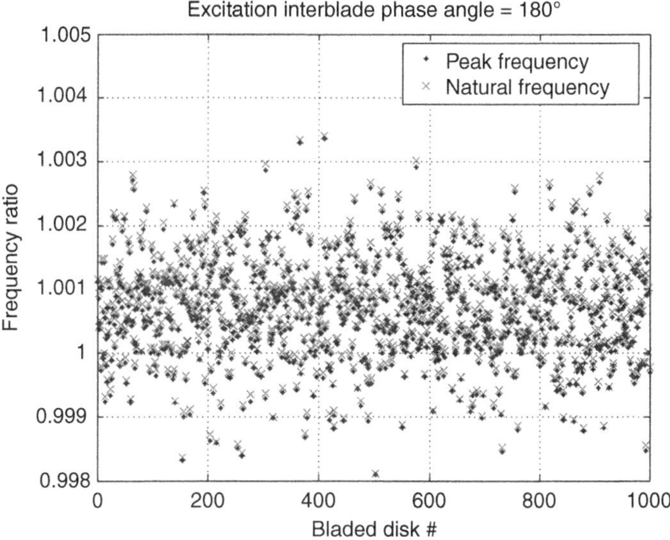

**Figure 2.2.3.**  Excitation frequency for the peak maximum amplitude $I_4$ ($k_c = 4543$, $c = 0.138$, $ns = 1000$, $nog = 10000$, $\omega_\ell = 0.95\omega_r$, $\omega_u = 1.05\omega_r$, $\ell = 5$ (180 degrees), no. of blades = 10, $\sigma_k = 2000$), frequency grid is fixed and based on $\omega_r$ of tuned system. Peak frequency is the frequency for $I_4$. Natural frequency is for mistuned disk (Sinha, 2008a).

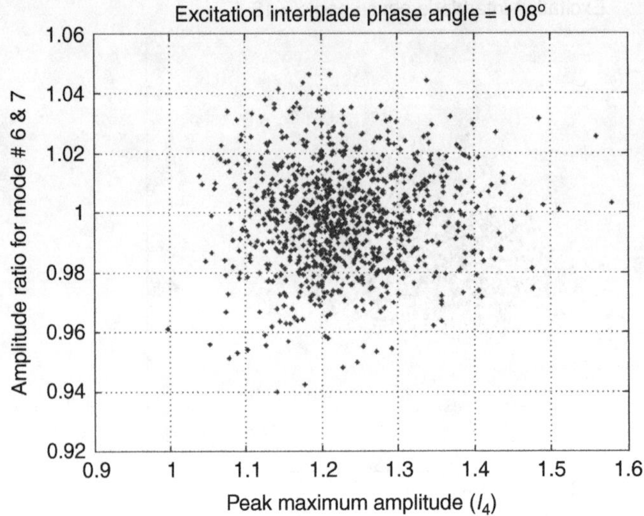



**Figure 2.2.4.**  Contribution of resonant modes to the peak maximum amplitude ($k_c = 4543$, $c = 0.138$, $ns = 1000$, $nog = 10000$, $\omega_\ell = 0.95\omega_r$, $\omega_u = 1.05\omega_r$, $\ell = 3$ (108 degrees), no. of blades = 10, $\sigma_k = 2000$), frequency grid is fixed and based on $\omega_r$ of tuned system (Sinha, 2008a).

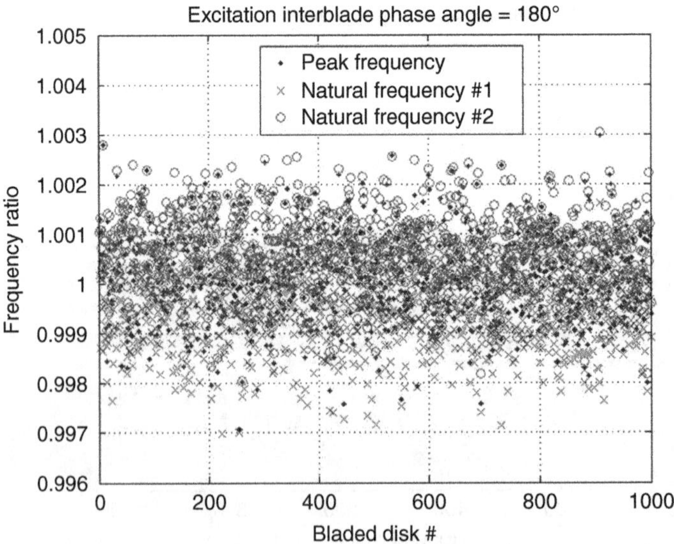



**Figure 2.2.5.**  Excitation frequency for the peak maximum amplitude $I_4$ ($k_c = 4543$, $c = 0.138$, $ns = 1000$, $nog = 10000$, $\omega_\ell = 0.95\omega_r$, $\omega_u = 1.05\omega_r$, $\ell = 3$ (108 degrees), no. of blades = 10, $\sigma_k = 2000$), frequency grid is fixed and based on $\omega_r$ of tuned system. Peak frequency is the frequency for $I_4$. Natural frequency # 1 and # 2 are for the mistuned disk (Sinha, 2008a).

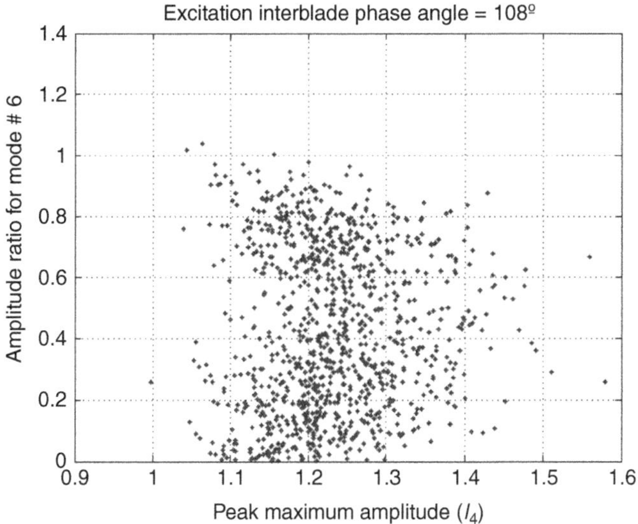



**Figure 2.2.6.** Contribution of one of the resonant modes to the peak maximum amplitude ($k_c = 4543$, $c = 0.138$, $ns = 1000$, $nog = 10000$, $\omega_\ell = 0.95\omega_r$, $\omega_u = 1.05\omega_r$, $\ell = 3$ (108 degrees), no. of blades = 10, $\sigma_k = 2000$), frequency grid is fixed and based on $\omega_r$ of tuned system (Sinha, 2008a).

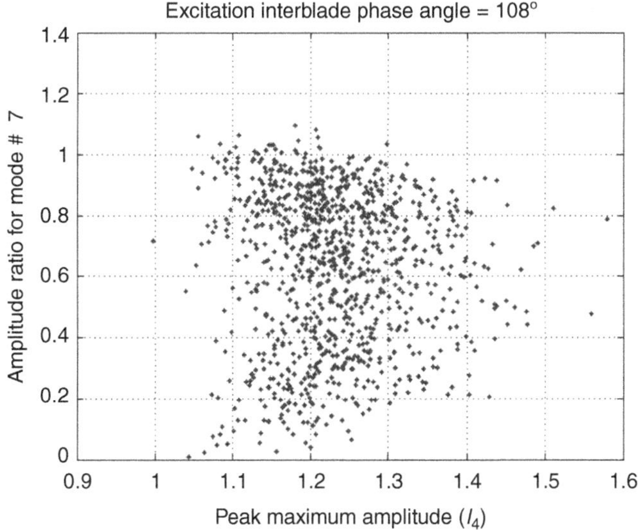



**Figure 2.2.7.** Contribution of one of the resonant modes to the peak maximum amplitude ($k_c = 4543$, $c = 0.138$, $ns = 1000$, $nog = 10000$, $\omega_\ell = 0.95\omega_r$, $\omega_u = 1.05\omega_r$, $\ell = 3$ (108 degrees), no. of blades = 10, $\sigma_k = 2000$), frequency grid is fixed and based on $\omega_r$ of tuned system (Sinha, 2008a).

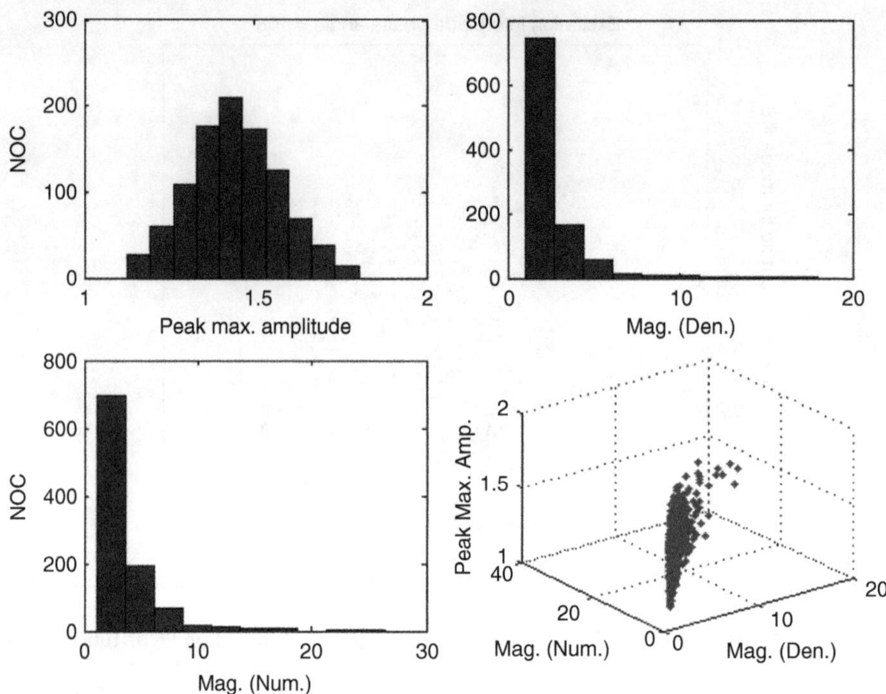

**Figure 2.2.8.** Numerators and denominators of peak maximum amplitude transfer functions ($k_c = 4543$, $c = 0.138$, $ns = 1000$, $nog = 1000$, $\omega_\ell = 0.95\omega_r$, $\omega_u = 1.05\omega_r$, $\ell = 5$ (180 degrees), no. of blades = 10, $\sigma_k = 2000$), frequency grid is fixed and based on $\omega_r$ of tuned system. Result is based on transfer function consisting of all modes (Sinha, 2008a).

mode shapes, it implies that mode shape changes contribute more toward amplitude amplification, which has also been reported by Kenyon, Griffin, and Feiner (2003) and Kenyon and Griffin (2003). As an example, poles and zeros of the tuned system and a mistuned bladed disk are listed in Tables 2.1.1 and 2.2.1 for $\ell = 1$, that is, excitation interblade phase angle = $36^0$. For the tuned system, all zeros are cancelled by poles, and the effective order of the tuned system is just two, which is true for any number of blades. With mistuning, there is no pole-zero cancellation and the order of the system is $2n$. While the imaginary parts appear in conjugate pairs, real parts are not identical for each pair, that is, zeros do not occur in complex conjugate pairs. If $\sigma + \iota\theta$ is a zero, then the corresponding factor in the numerator magnitude will be

$$\sqrt{\sigma^2 + (\omega - \theta)^2} \tag{2.2.55}$$

Note that $(\omega - \theta)^2$ will be small for all zeros with $\theta > 0$ when the coupling stiffness $k_c$ is small. As a result, $\sigma^2$ is the dominant term, and variations in $\sigma$ will lead to significant variations in amplitudes.

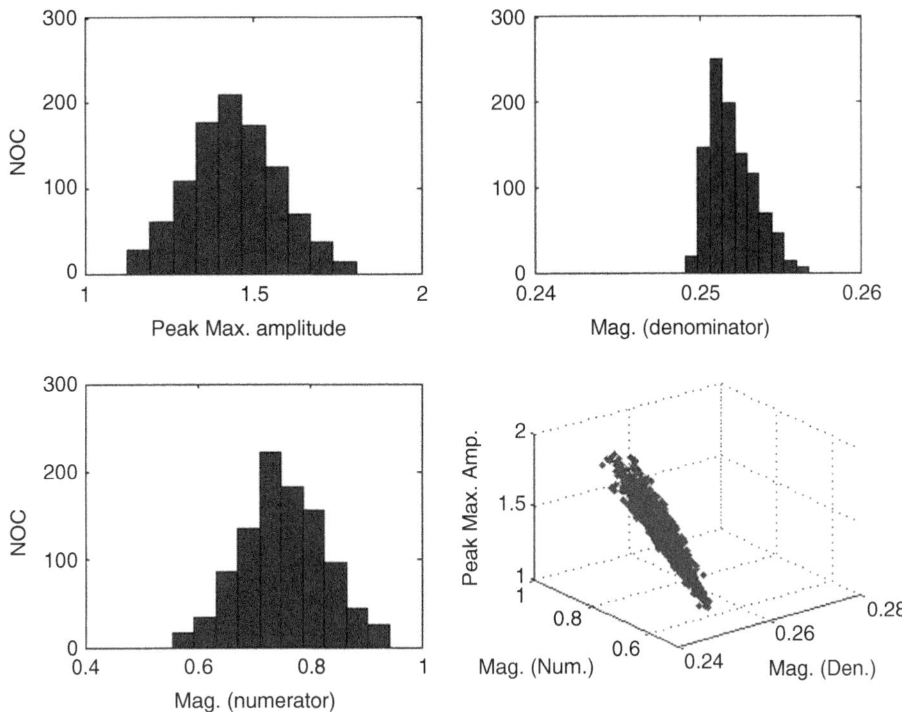

**Figure 2.2.9.** Numerators and denominators of peak maximum amplitude transfer function ($k_c = 4543$, $c = 0.138$, $ns = 1000$, $nog = 1000$, $\omega_\ell = 0.95\omega_r$, $\omega_u = 1.05\omega_r$, $\ell = 5$ (180 degrees), no. of blades = 10, $\sigma_k = 2000$), frequency grid is fixed and based on $\omega_r$ of tuned system. Results are from decomposition based on mistuned modes, and only considering the resonant mistuned mode (Sinha, 2008a).

It is also found that extreme values of numerator and denominator do not correspond to the maximum value of the peak maximum amplitude. In the example presented, numerator and denominator changed by factors of 7.23 and 4 for the peak maximum amplitude, respectively.

Next, numerator and denominator of the transfer function are obtained on the basis of resonant mistuned modes, that is, Equation (2.2.48) for unrepeated mode, and Equation (2.2.50) for repeated modes. In Figure 2.2.9, results are presented for the $180^0$ interblade phase angle. Magnitudes of numerators and denominators have been scaled with respect to their tuned values. The variations in denominator or poles are quite small whereas those in the numerator or zeros can be as high as 40 percent. Here, the peak maximum amplitude corresponds to a high value (0.9387) of the numerator magnitude, which is slightly less than its maximum value (0.9416). The corresponding denominator magnitude (0.2530) is found to be exactly the average of its maximum (0.2568) and minimum (0.2491) values.

Next, magnitudes of numerator and denominator are computed using Taylor series expansions (see Equations (2.2.53) and (2.2.54)). Partial derivatives of $q_i$ ($u_i$ or $v_i$) are computed by finite difference schemes as follows:

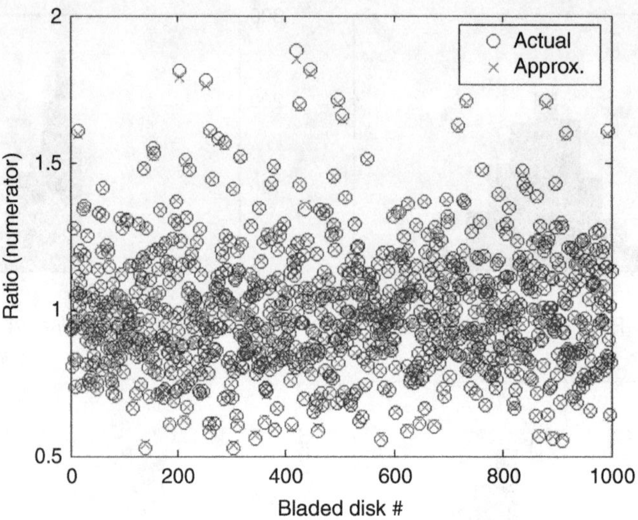


**Figure 2.2.10.** Prediction of numerator of the transfer function using Taylor Series ($k_c = 45430$, $c = 1.38$, $ns = 1000$, $\omega = \omega_r$, $\ell = 5$ (180 degrees), no. of blades = 10, $\sigma_k = 4000$), no modal decomposition (Sinha, 2008a).

$$\left( \frac{\partial q_i}{\partial k_j} \right) = \frac{q_i(k_t + \delta k_j) - q_i(k_t - \delta k_j)}{2\delta k_j} \tag{2.2.56}$$

$$\left( \frac{\partial^2 q_i}{\partial k_j^2} \right) = \frac{q_i(k_t + \delta k_j) + q_i(k_t - \delta k_j) - 2q_i(k_t)}{\delta k_j^2} \tag{2.2.57}$$

$$\left( \frac{\partial^2 q_i}{\partial k_\ell \partial k_j} \right) = \frac{q_i(k_t + \delta k_\ell, k_t) + q_i(k_t, k_t + \delta k_j) - q_i(k_t - \delta k_\ell, k_t) - q_i(k_t, k_t - \delta k_j)}{4\delta k_\ell \delta k_j} \tag{2.2.58}$$

For the purpose of computation of partial derivatives, $\delta k_\ell = \delta k_j = 1000 N/m$. The results from Equations (2.2.53) and (2.2.54) are compared to actual values with the excitation frequency being equal to the resonant frequency of the tuned system. When $k_c = 45430$N/m, and $c_t = 1.38$N-sec./meter, approximations (2.2.53) and (2.2.54) are quite valid up to mistuning standard deviation $\sigma_k = 4000$ N/m (see Figures 2.2.10 and 2.2.11). However, for $k_c = 4543$N/m, and $c_t = 1.38$N-sec./meter, approximations (2.2.53) and (2.2.54) are not found to be accurate.

### Nature of Statistical Distribution of Peak Maximum Amplitude

Modal parameters of each tuned blade are again chosen as $m_t = 0.0114 kg$. and $k_t = 430000 N/m$. The number of blades $n = 10$, and the engine-order excitation $\ell = 3$. Using MATLAB (1992), numerical simulations are performed for following values of standard deviation of mistuning $\sigma_k$: 1,000, 4,000, 7,000, 10,000, 13,000, 16,000, 19,000, and 22,000 N/m, at three values (low, intermediate, and high) of

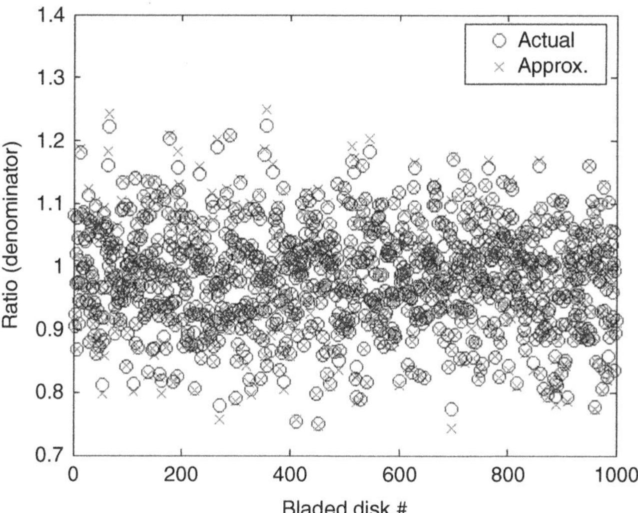



**Figure 2.2.11.** Prediction of the denominator of the transfer function using Taylor Series ($k_c = 45430$, $c = 1.38$, $ns = 1000$, $\omega = \omega_r$, $\ell = 5$ (180 degrees), no. of blades = 10, $\sigma_k = 4000$), no modal decomposition (Sinha, 2008a).

coupling stiffness $k_c$ = 4543, 45430, and 454300 N/m. The damping coefficient $c_t$ is chosen to be 0.138 N-sec/m, which corresponds to about 0.1 percent damping ratio. The standard deviation $\sigma_k$ = 10,000 N/m represents about 1 percent mistuning in terms of ratio of standard deviation and mean of blades' natural frequencies.

The distributions of the peak maximum amplitude, $B = I_4$, Equation (2.2.22), obtained from 10,000 simulations in each case, are presented in Figures 2.2.12– 2.2.15. On the top of each subplot, there are two numbers. The left number is the value of $k_c$ in N/m, and the right number is the standard deviation of modal stiffness $\sigma_k$ in N/m. Looking at these plots, it is obvious that the nature of distribution varies a great deal as a function of the coupling stiffness $k_c$ and the standard deviation of modal stiffness $\sigma_k$. It seems to be almost Gaussian for some cases, for example, $k_c = 4543 N/m$ and $\sigma_k = 1,000$ N/m. Some of the distributions can have two peaks, for example, $k_c = 454300 N/m$ and $\sigma_k = 4,000$ N/m., or long tails, for example, $k_c = 454300 N/m$ and $\sigma_k = 22,000$ N/m.

A two-parameter Weibull probability density function for the peak maximum amplitude $B = I_4$ is described (Statistics Toolbox, 2002) as

$$g(B) = \alpha\beta B^{\beta-1} \exp(-\alpha B^\beta)\psi \qquad (2.2.59a)$$

where $\psi$ is 1 for $B \geq 0$ and 0 otherwise. Here, $\alpha$ and $\beta$ are constant parameters. Sinha (2005) has found that none of distributions in Figures 2.2.12–2.2.15 is represented by a two-parameter Weibull distribution.

A three-parameter Weibull probability density function is described by introducing an additional constant parameter $\gamma$ as follows:

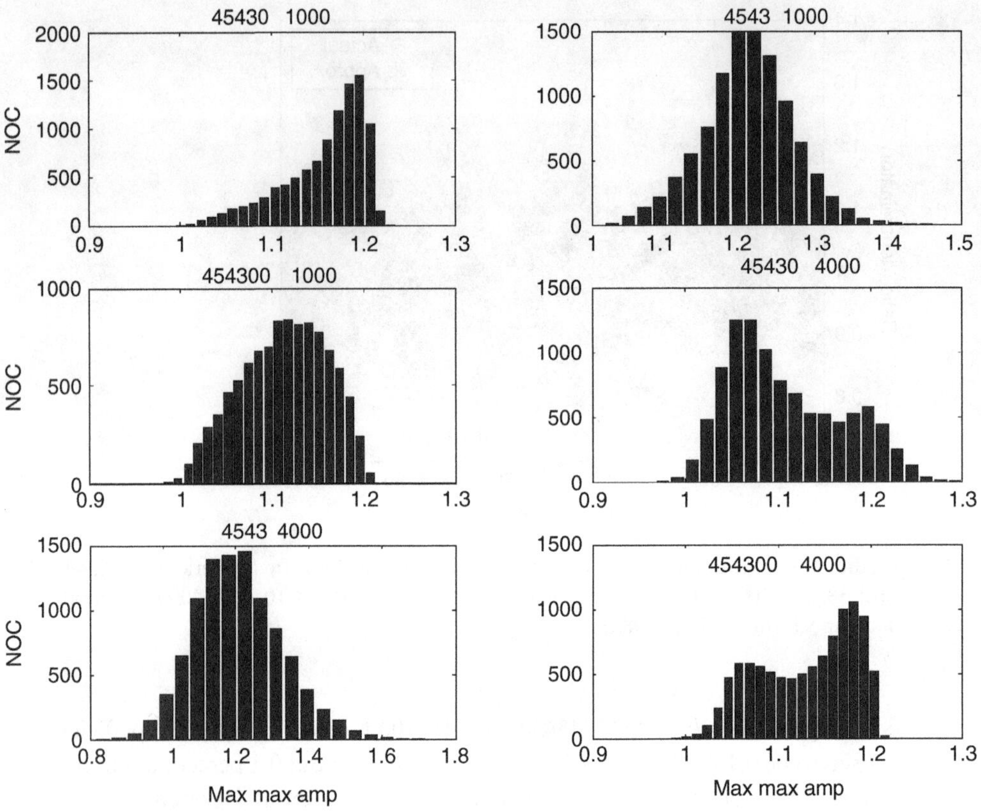


**Figure 2.2.12.** Distribution of B, peak maximum amplitude (Sinha, 2006b). (Vertical axes represent number of occurrences.)

$$g(B) = \alpha\beta(\gamma - B)^{\beta-1} \exp(-\alpha(\gamma - B)^{\beta})\psi \qquad (2.2.59b)$$

The probability distribution function corresponding to Equation (2.2.59b) is

$$F(B) = \int_{0}^{B} g(B)dB = \exp(-\alpha(\gamma - B)^{\beta}) - \exp(-\alpha\gamma^{\beta}) \qquad (2.2.60)$$

For typical values of $\alpha$, $\beta$ and $\gamma$,

$$\exp(-\alpha\gamma^{\beta}) \approx 0 \qquad (2.2.61)$$

Therefore, from Equation (2.2.60),

$$F(B) = \int_{0}^{B} g(B)dB = \exp(-\alpha(\gamma - B)^{\beta}) \qquad (2.2.62)$$

which implies that $F(\gamma) = 1$ and the maximum value of the peak maximum amplitude $B$ is $\gamma$.

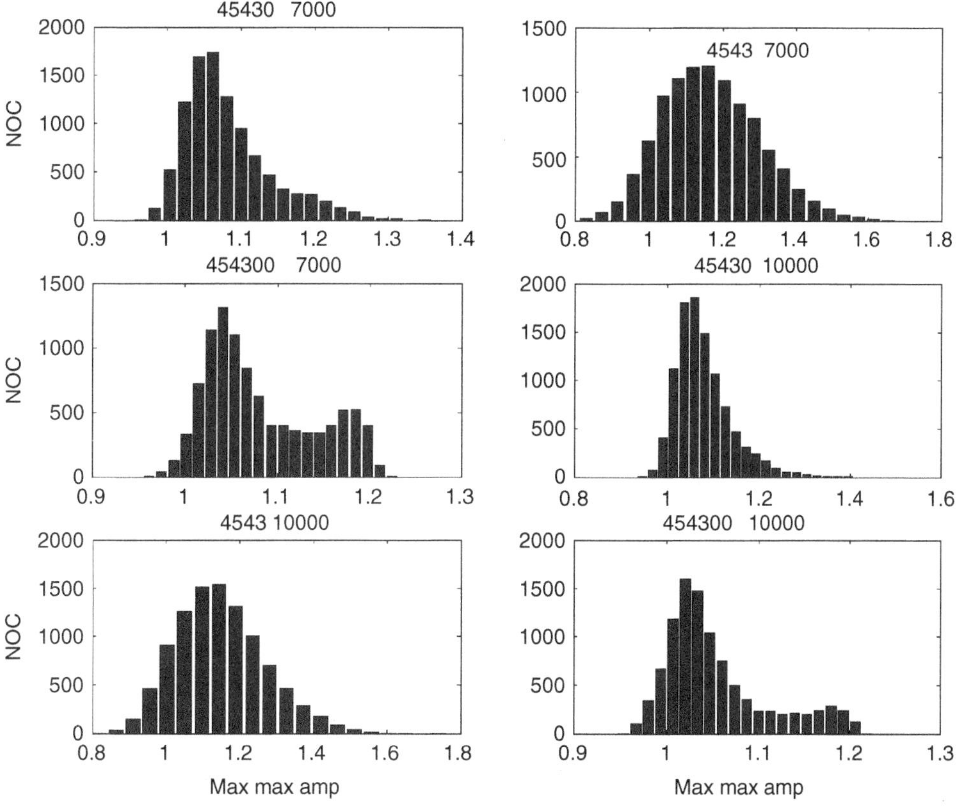

**Figure 2.2.13.** Distribution of B, peak maximum amplitude (Sinha, 2006b). (Vertical axes represent number of occurrences.)

For the three-parameter probability density function (2.2.59), Mignolet, Hu, and Jadic (2000) have shown that

$$E((\gamma - B)^k) = \alpha^{-k/\beta}\Gamma\left(1+\frac{k}{\beta}\right); \quad k = 1,2,3,\ldots \tag{2.2.63}$$

where $E(.)$ and $\Gamma(.)$ are expected value and gamma function, respectively. With $k = 1$, 2, and 3, Equation (2.2.63) leads to

$$\gamma - E(B) = \alpha^{-1/\beta}\Gamma\left(1+\frac{1}{\beta}\right) \tag{2.2.64}$$

$$\gamma^2 - 2\gamma E(B) + E(B^2) = \alpha^{-2/\beta}\Gamma\left(1+\frac{2}{\beta}\right) \tag{2.2.65}$$

$$\gamma^3 - E(B^3) - 3\gamma^2 E(B) + 3\gamma E(B^2) = \alpha^{-3/\beta}\Gamma\left(1+\frac{3}{\beta}\right) \tag{2.2.66}$$

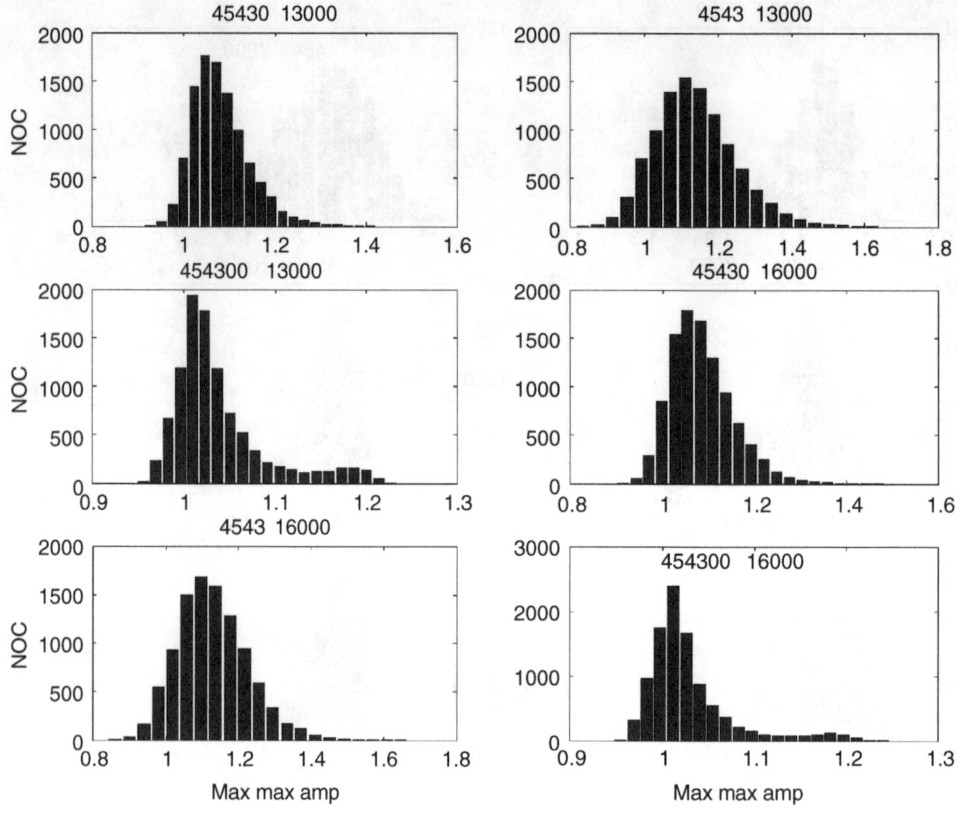

**Figure 2.2.14.** Distribution of B, peak maximum amplitude (Sinha, 2006b). (Vertical axes represent number of occurrences.)

Using $E(B)$, $E(B^2)$, and $E(B^3)$ from Monte Carlo simulations, nonlinear Equations (2.2.64)–(2.2.66) are numerically solved to determine Weibull parameters: $\alpha$, $\beta$, and $\gamma$.

The distributions of peak maximum amplitudes (see Figures 2.2.12–2.2.15) are examined on Weibull plots, see Figures 2.2.16–2.2.19 using MATLAB command "weibplot" (Statistics Toolbox, 2002). On the top of each subplot, there are two numbers. The left number is the value of $k_c$ in N/m, and the right number is the standard deviation of modal stiffness $\sigma_k$ in N/m. If the distribution is three-parameter Weibull, it will show up as a straight line on the Weibull plot. Non-Weibull distributions introduce curvatures in Weibull plots. Examining Figures 2.2.16–2.2.19, it is concluded that distributions of peak maximum amplitudes are three-parameter Weibull in some cases, but not always. It should be noted that extreme values of $B$ are mapped to the left end of each subplot in Figures 2.2.16–2.2.19. This conclusion is also confirmed by direct comparison between Monte Carlo distributions and three-parameter Weibull distributions (2.2.59), for example, Figures 2.2.20 and 2.2.21 where the range of $B$ is divided into 20 equal parts, and the number of

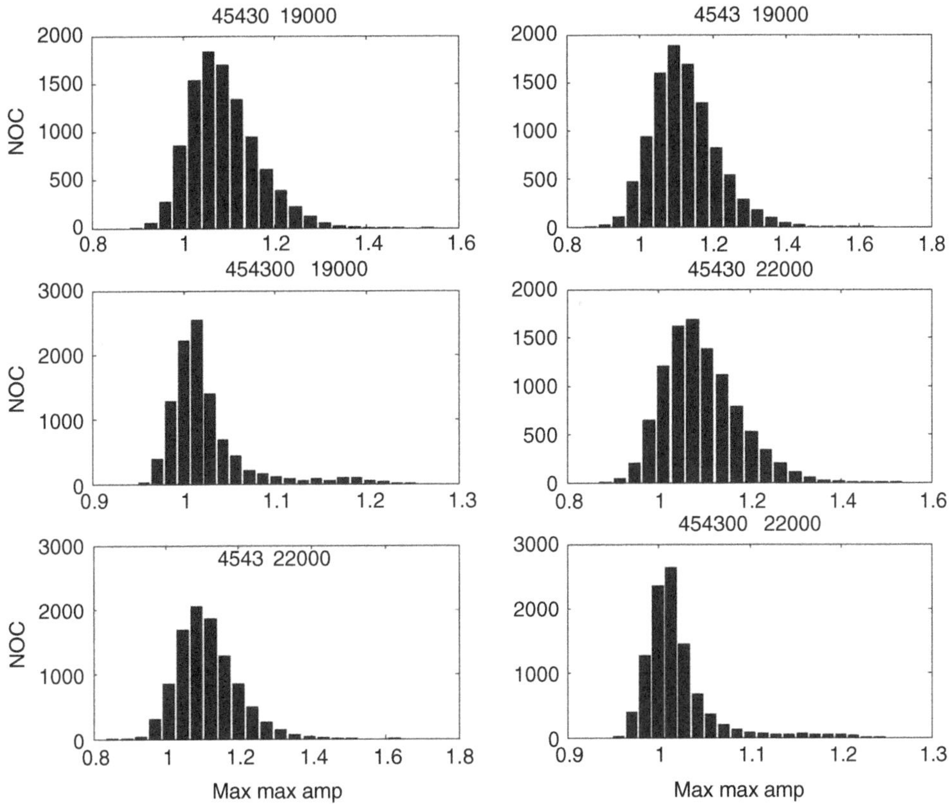


**Figure 2.2.15.** Distribution of B, peak maximum amplitude (Sinha, 2006b). (Vertical axes represent NOCs.)

occurrences of amplitudes is counted in each part. The first value of NOC corresponds to first of the twenty equal parts and so on.

For a low coupling stiffness, the distribution of the peak maximum amplitude can be treated as a three-parameter Weibull as shown by Castanier and Pierre (1997). In fact, Mignolet, Rivas-Guerra, and LaBorde (1999) also found distributions to be three-parameter Weibull for a low $k_c = 2000$ and a low-intermediate $k_c = 20,000$ N/m. However, they did not show any case for which the distribution is not three-parameter Weibull.

## 2.3　Maximum Amplitude of Vibration of a Mistuned Bladed Disk

### 2.3.1　Whitehead's Classical Result

Whitehead (1966, 1998) derived an important expression for the maximum amplitude of vibration of a mistuned bladed disk with single mode representation of each blade. Although his result was derived for structural and aerodynamic couplings

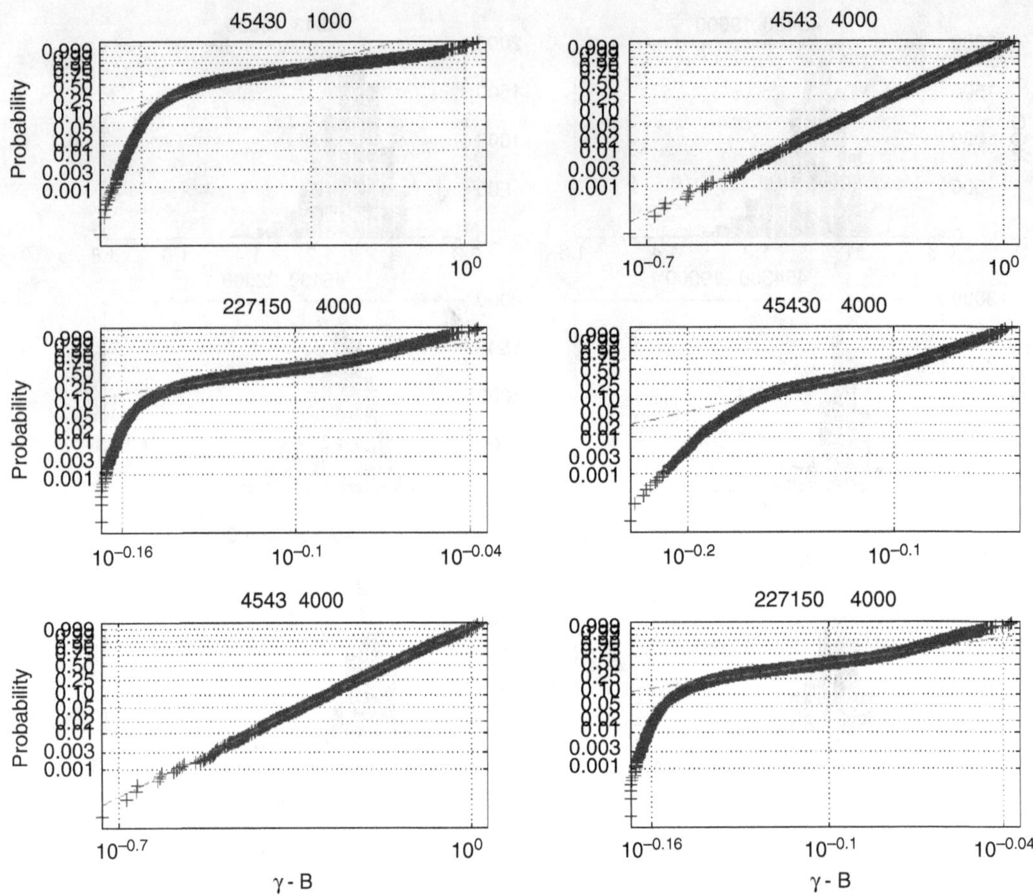



**Figure 2.2.16.** Weibull plots (Sinha, 2006b) for distributions of peak maximum amplitudes (Model number 1, $\sigma_k$: 1,000 and 4,000 N/m).

among all blades, derivation here is provided for the model shown in Figure 2.1.1. Equation (2.2.3) is rewritten in the form used by Whitehead (1998):

$$-m_t\omega^2 a_i = -k_t(1+\gamma_i)a_i + k_t\mu(-2+a_{i+1}+\mu a_{i-1}) - j(k_t\delta/\pi)a_i + k_t h_{\ell i}; \quad i = 1,2,\cdots,n \tag{2.3.1}$$

where

$$\gamma_i = \frac{\delta k_i}{k_t}, \quad \mu = \frac{k_c}{k_t}, \quad \delta = \frac{c_t\omega\pi}{k_t}; \quad \mathbf{h}_\ell = \frac{\mathbf{p}_\ell}{k_t}; \quad j = \sqrt{-1} \tag{2.3.2}$$

and $a_i$ is the complex amplitude of each blade. Define

$$\omega_0^2 = \frac{k_t}{m_t} \quad \text{and} \quad p = \frac{\omega^2}{\omega_0^2} - 1 \tag{2.3.3–2.3.4}$$

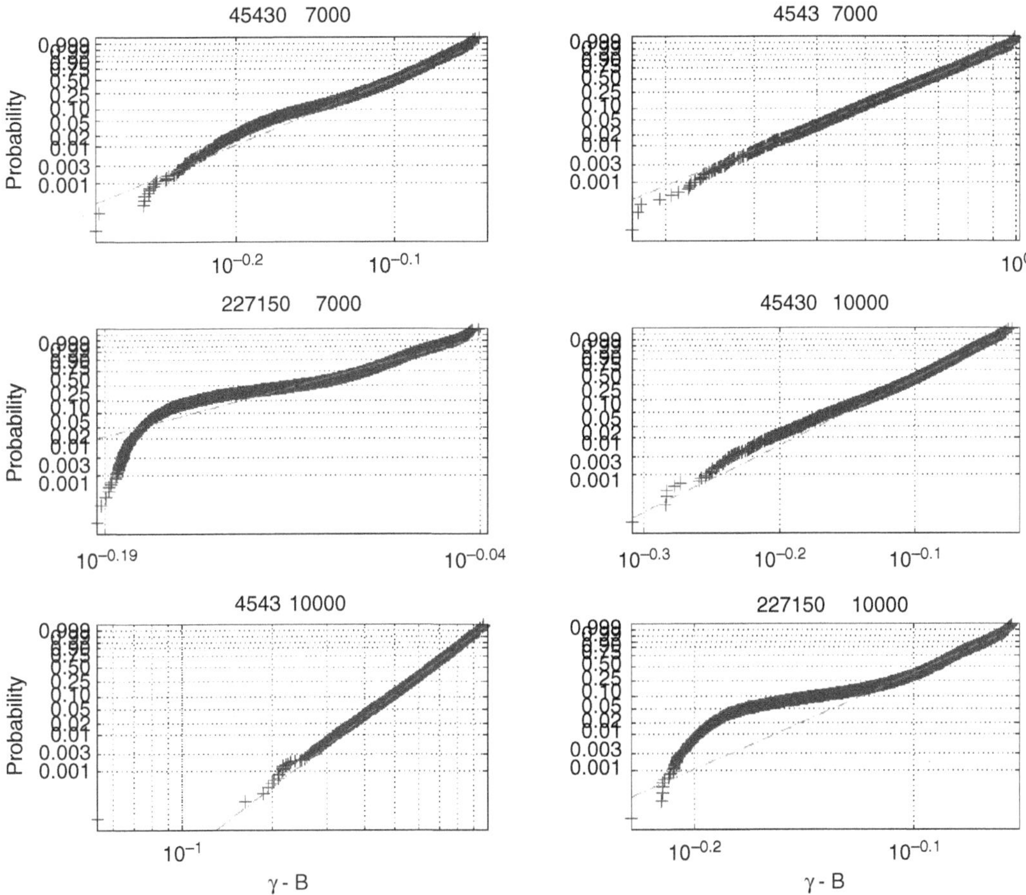

**Figure 2.2.17.** Weibull plots (Sinha, 2006b) for distributions of peak maximum amplitudes ($\sigma_k$: 7,000 and 10,000 N/m).

Dividing Equation (2.3.1) by $k_t$ and representing these equations in matrix form:

$$(\Gamma + j\delta/\pi I_n - pI_n - E_t)\mathbf{a} = \mathbf{h}_\ell \qquad (2.3.5)$$

where $\Gamma$ is a diagonal matrix with elements $\gamma_i$, that is,

$$\Gamma = diag\left[\gamma_1 \quad \gamma_2 \quad \cdots \quad \gamma_n\right] \qquad (2.3.6)$$

and

$$E_t = \frac{1}{k_t} K_t \qquad (2.3.7)$$

The matrix $K_t$ is defined by Equation (1.1.8). Define vectors $\mathbf{b}$ and $\mathbf{g}$ as follows:

$$\mathbf{a} = \Phi\mathbf{b} \quad \text{and} \quad \mathbf{h}_\ell = \Phi\mathbf{g} \qquad (2.3.8\text{–}2.3.9)$$

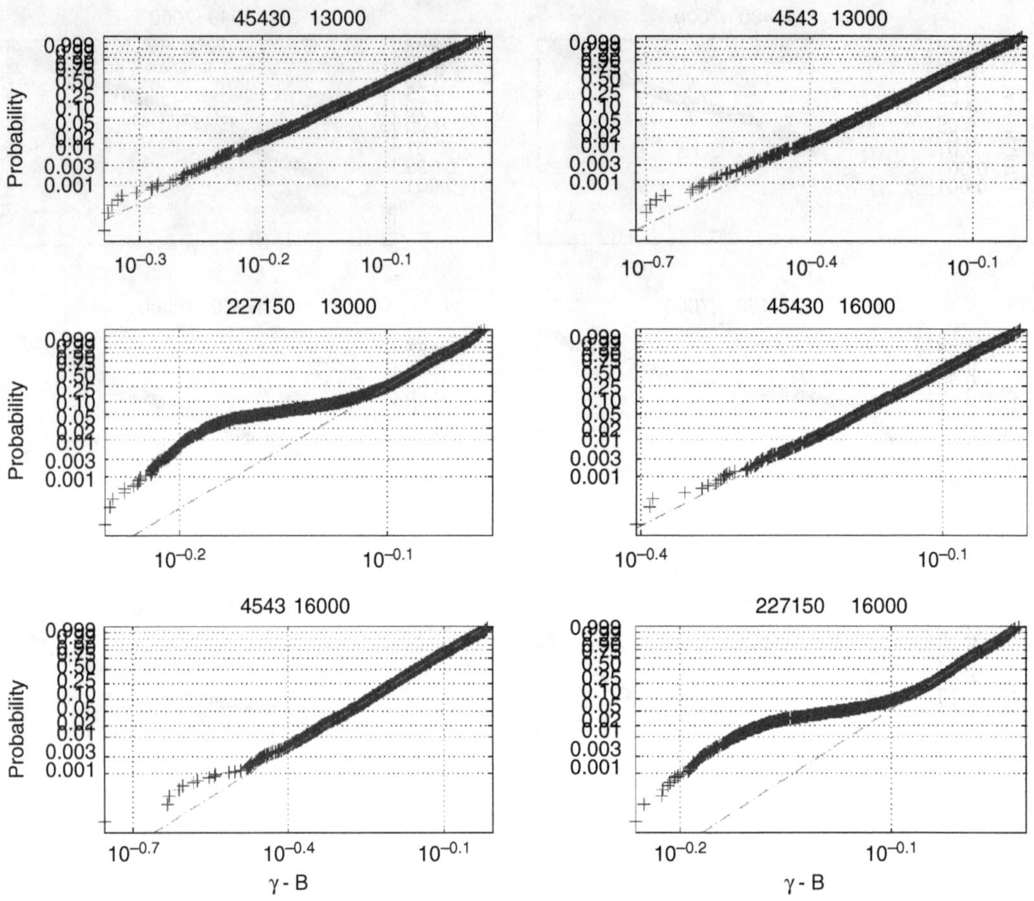


**Figure 2.2.18.** Weibull plots (Sinha, 2006b) for distributions of peak maximum amplitudes ($\sigma_k$: 13,000 and 16,000 N/m).

For the engine excitation order $= \ell$,

$$g_r = 0 \quad \text{for} \quad r \neq \ell \tag{2.3.10}$$

Substituting Equations (2.3.8–2.3.9) into Equation (2.3.5) and premultiplying Equation (2.3.5) by $\Phi^{-1}$,

$$(\Phi^{-1}\Gamma\Phi + j\delta / \pi I - pI - \Phi^{-1}E_t\Phi)\mathbf{b} = \mathbf{g} \tag{2.3.11}$$

Note that

$$\Phi^{-1}E_t\Phi = diag[\varepsilon_{11} \quad \varepsilon_{22} \quad , \quad . \quad \varepsilon_{nn}] \tag{2.3.12}$$

$$\varepsilon_{rr} = \frac{\omega_{fr}^2}{\omega_0^2} \tag{2.3.13}$$

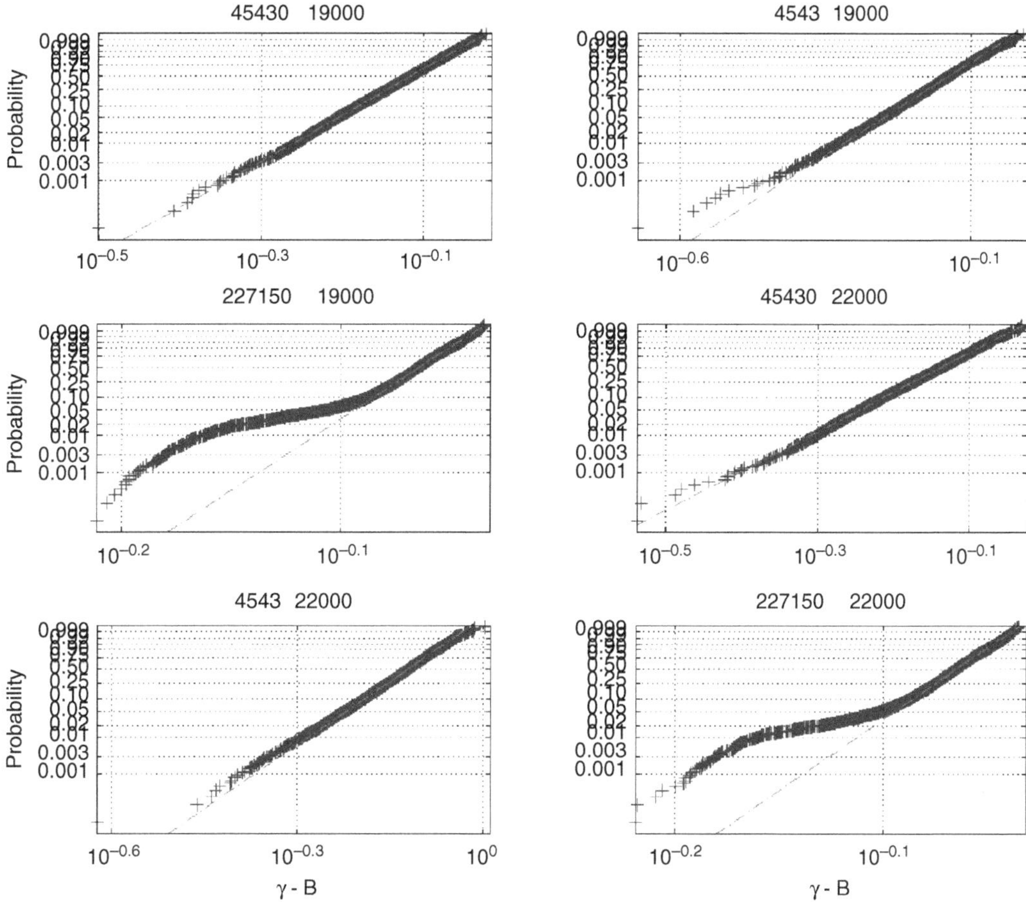


**Figure 2.2.19.**  Weibull plots (Sinha, 2006b) for distributions of peak maximum amplitudes ($\sigma_k$: 19,000 and 22,000 N/m).

and $\omega_{fr}$ is the natural frequency of a perfectly tuned system, Equation (1.2.9).

### Perfectly Tuned Case

In this case, $\Gamma = 0$ and Equation (2.3.11) reduces to

$$(j\delta/\pi I - pI - \Phi^{-1}E_t\Phi)\mathbf{b} = \mathbf{g} \tag{2.3.14}$$

For the element number $r$,

$$(j\delta/\pi - p - \varepsilon_{rr})b_r = g_r \tag{2.3.15}$$

At resonance for the engine excitation order $= \ell$, $p - \varepsilon_{\ell\ell} = 0$. Therefore,

$$b_\ell = b_t = \frac{g_\ell}{j\delta/\pi} \quad \text{and} \quad b_r = 0 \text{ for } r \neq \ell \tag{2.3.16}$$

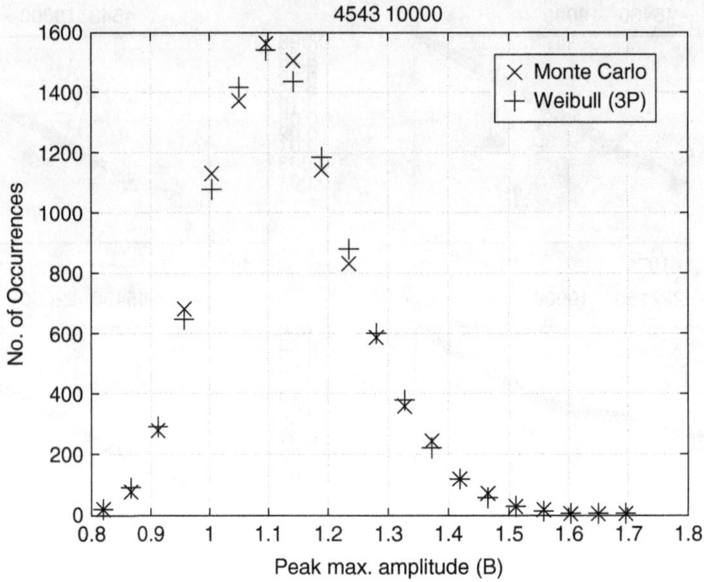



**Figure 2.2.20.** Direct comparison (Sinha, 2006b) between Monte Carlo and three-parameter Weibull distributions ($k_c = 4543N \, / \, m$, $\sigma_k = 10,000N \, / \, m$).

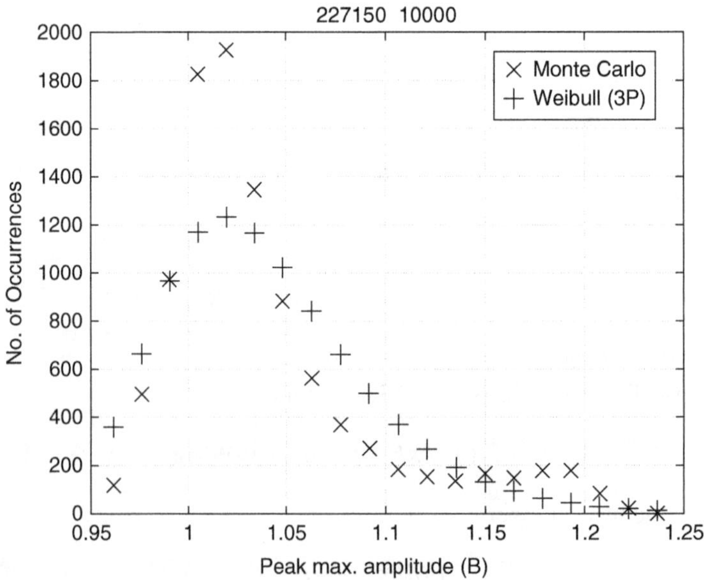



**Figure 2.2.21.** Direct comparison (Sinha, 2006b) between Monte Carlo and three-parameter Weibull distributions ($k_c = 227150N \, / \, m$, $\sigma_k = 10,000N \, / \, m$).

### Mistuned Case ($\Gamma \neq 0$)

Premultiplying both sides of Equation (2.3.11) by $\mathbf{b}^H$,

$$\mathbf{b}^H(\Phi^{-1}\Gamma\Phi + j\delta/\pi I - pI - \Phi^{-1}E_t\Phi)\mathbf{b} = \mathbf{b}^H\mathbf{g} = b_\ell^* g_\ell \tag{2.3.17}$$

Equating imaginary part on both sides,

$$(\delta/\pi)\sum_{r=1}^{n}\left|b_r\right|^2 = -\operatorname{Im} ag(b_\ell)g_\ell \tag{2.3.18}$$

or,

$$\sum_{r=1}^{n}\left|b_r\right|^2 = -\operatorname{Im} ag(b_\ell)\frac{\pi g_\ell}{\delta} = -\operatorname{Im} ag(b_\ell)\left|b_t\right| \tag{2.3.19}$$

or,

$$\sum_{r=1}^{n}\frac{\left|b_r\right|^2}{\left|b_t\right|^2} = -\frac{\operatorname{Im} ag(b_\ell)}{\left|b_t\right|}; \tag{2.3.20}$$

Define,

$$z_r = \frac{\left|b_r\right|}{\left|b_t\right|} \tag{2.3.21}$$

From Equation (2.3.20)

$$\sum_{r=1}^{n} z_r^2 \leq z_\ell \tag{2.3.22}$$

The maximum amplitude will occur when all traveling waves are in phase. Without any loss of generality, invoking this condition for the blade number 1, Equations (2.3.8) and (2.3.21) yield

$$\frac{a_{\max}}{\left|b_t\right|} = \sum_{r=0}^{n-1} z_r \tag{2.3.23}$$

Now, Equation (2.3.23) is maximized subject to the constraint (2.3.22) with equality sign. Hence, the objective function to be maximized is

$$J = \sum_{r=0}^{n-1} z_r + \lambda\left(\sum_{r=0}^{n-1} z_r^2 - z_\ell\right) \tag{2.3.24}$$

where $\lambda$ is a Lagrange multiplier. Necessary conditions for optimality yield

$$\frac{\partial J}{\partial z_r} = 1 + 2z_r\lambda = 0; \quad r \neq \ell, \ \Rightarrow z_r = -\frac{1}{2\lambda} \tag{2.3.25}$$

$$\frac{\partial J}{\partial z_\ell} = 1 + 2z_\ell \lambda - \lambda = 0; \quad \Rightarrow z_\ell = -\frac{1}{2\lambda} + \frac{1}{2} \tag{2.3.26}$$

$$\frac{\partial J}{\partial \lambda} = \sum_{r=0}^{n-1} z_r^2 - z_\ell = 0 \tag{2.3.27}$$

Substituting Equations (2.3.25) and (2.3.26) into Equation (2.3.27),

$$\sum_{r=0}^{n-1} z_r^2 - z_\ell = \frac{n-1}{4\lambda^2} + \frac{1}{4} + \frac{1}{4\lambda^2} - \frac{1}{2\lambda} + \frac{1}{2\lambda} - \frac{1}{2} = \frac{n}{4\lambda^2} - \frac{1}{4} = 0 \tag{2.3.28}$$

From Equation (2.3.28),

$$\lambda = -\sqrt{n} \tag{2.3.29}$$

From Equations (2.3.25) and (2.3.26),

$$z_r = \frac{1}{2\sqrt{n}} \quad \text{and} \quad z_\ell = \frac{1}{2\sqrt{n}} + \frac{1}{2} \tag{2.3.30}$$

From Equation (2.3.23), the maximum amplitude amplification is given by

$$\frac{a_{\max}}{|b_\ell|} = \sum_{r=0}^{n-1} z_r = \frac{1}{2} + \frac{n}{2\sqrt{n}} = \frac{1}{2}(1 + \sqrt{n}) \tag{2.3.31}$$

Equation (2.3.31) indicates that the maximum amplitude amplification due to mistuning is $(1 + \sqrt{n})/2$. However, it must be noted that this result has been derived for a model with a single mode of vibration for each blade, identical damping for each blade, and the excitation frequency equal to a natural frequency of a tuned disk. Recently, Kenyon and Griffin (2003), Rivas-Guerra and Mignolet (2003), Xiao, Rivas-Guerra, and Mignolet (2004), and Chan and Ewins (2011) have also analytically derived the expressions for the maximum possible amplitude by considering distortion of structural modes, damping variations, and multiple degree of freedom for each sector as well. It has been shown that the maximum amplitude amplification due to mistuning can exceed $(1 + \sqrt{n})/2$.

## 2.3.2    Numerical Approach

### Computation through Infinity Norm

For engine-order excitation (2.1.7), the steady-state response can be written as

$$\mathbf{x}(t) = \mathbf{a}\, e^{\iota \omega t}; \quad \iota = \sqrt{-1} \tag{2.3.32}$$

Substituting Equation (2.3.32) into Equation (2.1.2),

$$H\mathbf{a} = \mathbf{p}_\ell \tag{2.3.33}$$

where

$$H = K - \omega^2 M + \iota\omega C, \tag{2.3.34}$$

For nonzero damping, the inverse of matrix $H$ exists for all frequencies $\omega$. From Equation (2.3.33),

$$\mathbf{a} = H^{-1}\mathbf{p}_\ell \tag{2.3.35}$$

Using infinity norm (Vidyasagar, 1993, Sinha, 2007), an upper bound of the maximum amplitude at a fixed frequency $\omega$ is derived as

$$\|\mathbf{a}\|_\infty \leq \|H^{-1}\|_\infty \|\mathbf{p}_\ell\|_\infty \tag{2.3.36}$$

where

$$\|\mathbf{a}\|_\infty = \max_i |a_i| \tag{2.3.37}$$

$$\|H^{-1}\|_\infty = \max_i \sum_{\ell=1}^n |(H^{-1})_{i\ell}| \tag{2.3.38}$$

Without any loss of generality, $\|\mathbf{p}_\ell\|_\infty = 1$. Therefore, an upper bound of the maximum amplitude at a specified frequency $\omega$ is

$$\|\mathbf{a}\|_\infty \leq \|H^{-1}\|_\infty \text{ for any engine-order excitation} \tag{2.3.39}.$$

Assuming that the mistuning is caused by variations in stiffnesses alone, define

$$\delta\mathbf{k} = \begin{bmatrix} \delta k_1 & \delta k_2 & \cdots & \delta k_n \end{bmatrix}^T \tag{2.3.40}$$

First the following optimization problem is solved:

$$\max_{\delta\mathbf{k}} \|H^{-1}\|_\infty \tag{2.3.41}$$

The results obtained for all engine-order excitations with the coupling stiffness $k_c = 45430 \, N/m$ have been presented by Sinha (1997). The maximum amplification factor, Table 2.3.1, is close to that predicted by Whitehead (see Equation (2.3.31)). The corresponding values of mistuning parameters are presented in Table 2.3.2. Some of the variations in stiffnesses are much larger compared to typical values found in practice. Typically, standard deviations for variations in stiffness is around 1 percent, which translates to $\sigma_k = 10000 \, N/m$. Therefore, the constrained optimization problem is solved as:

$$\max_{\delta\mathbf{k}} \|H^{-1}\|_\infty \tag{2.3.42}$$

**Table 2.3.1.** Maximum amplification factor ($k_c = 45430\,N\,/\,meter$, $n=10$) (Sinha, 1997)

|  | $c_t = 0.138\text{Ns/m}$ | $c_t = 1.38\text{Ns/m}$ |
|---|---|---|
| $r = 0$ | 2.0804 | 2.0811 |
| $r = 1$ | 2.0543 | 2.0811 |
| $r = 2$ | 2.0796 | 2.0789 |
| $r = 3$ | 2.0810 | 2.0810 |
| $r = 4$ | 2.0748 | 2.0811 |
| $r = 5$ | 2.0810 | 2.0811 |
| $r = 6$ | 2.0755 | 2.0811 |
| $r = 7$ | 2.0807 | 2.0810 |
| $r = 8$ | 2.0795 | 2.0789 |
| $r = 9$ | 2.0658 | 2.0811 |

**Table 2.3.2.** Mistuning parameters from unconstrained optimization (Sinha, 1997) ($r = 3, k_c = 45430\,N\,/\,meter$, $n=10$)

|  | $c_t = 0.138\text{Ns/m}$ | $c_t = 1.38\text{ Ns/m}$ |
|---|---|---|
| $\delta k_1$ | 111434 N/m | −69259 N/m |
| $\delta k_2$ | −33207 N/m | −31545 N/m |
| $\delta k_3$ | −29637 N/m | 42349 N/m |
| $\delta k_4$ | −21718 N/m | 51885 N/m |
| $\delta k_5$ | 56391 N/m | −32449 N/m |
| $\delta k_6$ | −23243 N/m | −27937 N/m |
| $\delta k_7$ | −31609 N/m | −25622 N/m |
| $\delta k_8$ | −28389 N/m | 117399 N/m |
| $\delta k_9$ | 122326 N/m | −6127 N/m |
| $\delta k_{10}$ | −6035 N/m | 121886 N/m |

subject to

$$-3\sigma_k \leq \delta k_j \leq 3\sigma_k; \quad j = 1, 2, \dots, n \tag{2.3.43}$$

Results obtained by Sinha (1997) are shown in Tables 2.3.3 and 2.3.4. The maximum amplification factor is slightly less than that predicted by Whitehead (see Equation (2.3.31)).

The computation of the maximum amplitude, $I_1$-$I_4$, Equations (2.2.19)–(2.2.22), is a $np$-hard problem (Rotea and D'Amato, 2002 , D'Amato and Rotea, 2005), that is, it is computationally intractable except with a small number of mistuning parameters. Petrov and Ewins (2003) have attempted to find the worst mistuning pattern for a large finite element model of a bladed disk assembly through numerical optimization.

**Table 2.3.3.** Maximum amplification factor ($k_c = 45430\,N\,/\,meter$, $n = 10$) from constrained optimization (Sinha, 1997)

|         | $c_t = 0.138$Ns/m | $c_t = 1.38$Ns/m |
|---------|-------------------|------------------|
| $r = 0$ | 1.7511            | 1.7338           |
| $r = 1$ | 1.8933            | 1.9219           |
| $r = 2$ | 1.8618            | 1.8862           |
| $r = 3$ | 1.8763            | 1.8871           |
| $r = 4$ | 1.8915            | 1.9282           |
| $r = 5$ | 1.7512            | 1.7591           |
| $r = 6$ | 1.8915            | 1.9282           |
| $r = 7$ | 1.8763            | 1.8871           |
| $r = 8$ | 1.8618            | 1.8862           |
| $r = 9$ | 1.8926            | 1.9219           |

**Table 2.3.4.** Mistuning parameters from constrained optimization ($r = 3, k_c = 45430\,N\,/\,meter$) (Sinha, 1997)

|                | $c_t = 0.138$Ns/m | $c_t = 1.38$ Ns/m |
|----------------|-------------------|-------------------|
| $\delta k_1$   | 30000 N/m         | −30000 N/m        |
| $\delta k_2$   | −30000 N/m        | 2385 N/m          |
| $\delta k_3$   | 30000 N/m         | −30000 N/m        |
| $\delta k_4$   | 10143 N/m         | 30000 N/m         |
| $\delta k_5$   | −3266 N/m         | −30000N/m         |
| $\delta k_6$   | −30000 N/m        | 2386 N/m          |
| $\delta k_7$   | 6378 N/m          | −30000 N/m        |
| $\delta k_8$   | −30000 N/m        | 30000 N/m         |
| $\delta k_9$   | 30000 N/m         | 7941 N/m          |
| $\delta k_{10}$| 8327 N/m          | 30000 N/m         |

Thompson and Becus (1993) and Choi et al. (2003) have computed the best mistuning pattern, that is, minimum amplitude amplification, using simulated annealing and genetic algorithm, respectively. Jones (2004) has also presented an approach to determine the best mistuning pattern by minimizing modal forces. Rotea and D'Amato (2002) have presented an algorithm to compute upper and lower bounds for $I_3$, Equation (2.2.21), using robust control algorithm (Fan, Tits, and Doyle, 1991), which will be described next.

### Computation Using Linear Robust Control Theory

The governing differential equations of motion can be written as

$$M_\delta \ddot{\mathbf{x}} + C_\delta \dot{\mathbf{x}} + K_\delta \mathbf{x} = H\mathbf{f}_b(t) \tag{2.3.44}$$

where $M_\delta$, $C_\delta$, and $K_\delta$ are mass, damping, and stiffness matrices containing uncertain (mistuning) parameters $\delta_i$; $i = 1, 2, \ldots, np$ as follows:

$$M_\delta = M_t + \sum_{i=1}^{np} M_i \delta_i \tag{2.3.45}$$

$$C_\delta = C_t + \sum_{i=1}^{np} C_i \delta_i \tag{2.3.46}$$

$$K_\delta = K_t + \sum_{i=1}^{np} K_i \delta_i \tag{2.3.47}$$

Here, $M_t$, $C_t$, and $K_t$ are mass, damping, and stiffness matrices of a perfectly tuned system, respectively. The $nf \times 1$ vector $\mathbf{f}_b(t)$ represents $nf$ independent forcing functions. The matrix H is of dimension $n \times nf$. Matrices $M_i$, $C_i$, and $K_i$ are associated with uncertain parameter $\delta_i$. They can be factored as follows:

$$M_i = M_{Li} M_{Ri}; \; C_i = C_{Li} C_{Ri} \quad \text{and} \quad K_i = K_{Li} K_{Ri} \tag{2.3.48}$$

where $M_{Li}$, $C_{Li}$, and $K_{Li}$ are of dimension $n \times r_i$, and $M_{Ri}$, $C_{Ri}$, and $K_{Li}$ are of dimension $n \times r_i$, and $r_i$ is the maximum rank of matrices $M_i$, $C_i$, and $K_i$.

Let $\mathbf{y}(t)$ be a vector of those displacements for which maximum amplitudes are to be computed. In this case,

$$\mathbf{y}(t) = C_0 \mathbf{x}(t) \tag{2.3.49}$$

where $C_0$ is the selection matrix containing 0 or 1 as its elements. For sinusoidal excitation,

$$\mathbf{f}_b(t) = \mathbf{f}_{b0} e^{j\omega t}; \quad j = \sqrt{-1} \tag{2.3.50}$$

where $\mathbf{f}_{b0}$ is a complex vector. In steady state,

$$\mathbf{x}(t) = \mathbf{x}_0 e^{j\omega t} \tag{2.3.51}$$

and

$$\mathbf{y}(t) = \mathbf{y}_0 e^{j\omega t} \tag{2.3.52}$$

Substituting Equations (2.3.50) and (2.3.51) into (2.3.44), and equating coefficients of $e^{j\omega t}$ on both sides,

$$(P_t + \Delta P)\mathbf{x}_0 = H\mathbf{f}_{b0} \tag{2.3.53}$$

$$\mathbf{y}_0 = C_0 \mathbf{x}_0 \tag{2.3.54}$$

where

$$P_t = -\omega^2 M_t + j\omega C_t + K_t \tag{2.3.55}$$

$$\Delta P = \sum_{i=1}^{np} \delta_i P_i \tag{2.3.56}$$

$$P_i = -\omega^2 M_i + j\omega C_i + K_i \tag{2.3.57}$$

Using Equations (2.3.57) and (2.3.48)

$$P_i = P_i^L P_i^R \tag{2.3.58}$$

where

$$P_i^L = \left[ \; j\omega M_{Li} \quad (1+j)\sqrt{\frac{\omega}{2}}\, C_{Li} \quad K_{Li} \; \right] \tag{2.3.59}$$

$$P_i^R = \begin{bmatrix} j\omega M_{Ri} \\ (1+j)\sqrt{\dfrac{\omega}{2}}\, C_{Ri} \\ K_{Ri} \end{bmatrix} \tag{2.3.60}$$

Substituting Equation (2.3.58) into (2.3.56),

$$\Delta P = \sum_{i=1}^{np} P_i^L \delta_i P_i^R \tag{2.3.61}$$

Equation (2.3.61) can be expressed as

$$\Delta P = L\Delta(\delta)R \tag{2.3.62}$$

where

$$\Delta(\delta) = diag(\delta_1 I_{r_1} \quad \delta_2 I_{r_2} \quad \cdots \quad \delta_{np} I_{r_{np}}) \tag{2.3.63}$$

$$L = [P_1^L \quad P_2^L \quad \cdots \quad P_{np}^L] \tag{2.3.64}$$

$$R = [P_1^R \quad P_2^R \quad \cdots \quad P_{np}^R], \tag{2.3.65}$$

Let $G(s)$ be the transfer function relating output $\mathbf{y}(s)$ and forcing function $\mathbf{f}_b(s)$:

$$\mathbf{y}(s) = G(s)\mathbf{f}_b(s) \tag{2.3.66}$$

where

$$G(s) = C_0(s^2 M_\delta + sC_\delta + K_\delta)^{-1} H \tag{2.3.67}$$

Therefore,

$$\mathbf{y}_0 = G(j\omega)\mathbf{f}_{b0} \tag{2.3.68}$$

$$G(j\omega) = C_0(-\omega^2 M_\delta + j\omega C_\delta + K_\delta)^{-1} H \tag{2.3.69}$$

Using Equations (2.3.69) and (2.3.62)

$$G(j\omega) = C_0(P_t + L\Delta R)^{-1} H \tag{2.3.70}$$

Using Matrix Inversion Lemma (Sinha, 2007),

$$G(j\omega) = M_{22} + M_{21}\Delta(I - M_{11}\Delta)^{-1} M_{12} \tag{2.3.71}$$

where matrices $M_{11}$, $M_{12}$, $M_{21}$, and $M_{22}$ are given by

$$\begin{bmatrix} M_{11} & M_{12} \\ M_{21} & M_{22} \end{bmatrix} = \begin{bmatrix} R \\ C_0 \end{bmatrix} P_t^{-1} \begin{bmatrix} -L & H \end{bmatrix} \tag{2.3.72}$$

The worst-case frequency response at any $\omega$ is defined by

$$\gamma(\omega, \theta) = \max_{\delta} \ \overline{\sigma}(G(\omega, \delta)) \tag{2.3.73}$$

subject to

$$\delta_i \leq \theta, \ i = 1, 2, \ldots, np \tag{2.3.74}$$

where $\overline{\sigma}(G(\omega, \delta))$ is the maximum singular value of the complex matrix $G(\omega, \delta)$. The maximum singular value provides an upper bound of the 2-norm of the output amplitude vector (Sinha, 2007):

$$\frac{\|\mathbf{y}_0\|_2}{\|\mathbf{f}_{b0}\|_2} \leq \overline{\sigma}(G(\omega, \delta)) \tag{2.3.75}$$

where

$$\|\mathbf{y}_0\|_2 = \left[\sum_{i=1}^{n} |y_{0i}|^2\right]^{0.5} \quad \text{and} \quad \|\mathbf{f}_{b0}\|_2 = \left[\sum_{i=1}^{nf} |f_{b0i}|^2\right]^{0.5} \tag{2.3.76a, b}$$

In other words, maximum singular value is an upper bound on 2-norm, rather than on $\infty$-norm as described by Equation (2.3.37). However, describing the output as the displacement of a single node of a blade and having a single input, the maximum amplitude of that node at a frequency is same as the maximum singular value.

Rotea and D'Amato (2002) and D'Amato and Rotea (2005) computed the upper bound $\sqrt{\beta}$ on the worst-case frequency response $\gamma$, that is,

$$\gamma \leq \sqrt{\beta} \tag{2.3.77}$$

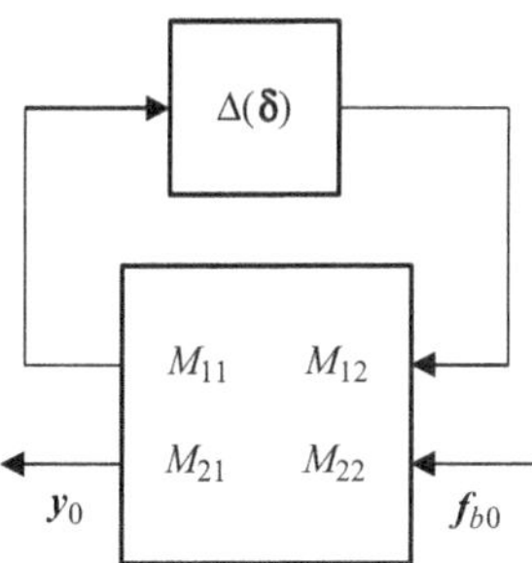

**Figure 2.3.1.**   Upper LFT.

where $\beta$ is obtained by solving the following Linear Matrix Inequality (LMI) problem:

$$\beta = \min q \tag{2.3.78}$$

$$X = X^H = diag(X_1, X_2, \ldots, X_n) \geq 0 \tag{2.3.79}$$

$$Y = -Y^H = diag(Y_1, Y_2, \ldots, Y_n) \tag{2.3.80}$$

$$\begin{bmatrix} \theta^{-2} X - M_{11}^H X M_{11} + Y M_{11} - M_{11}^H Y - M_{21}^H M_{21} & -(Y^H + M_{11}^H X) M_{12} - M_{21}^H M_{22} \\ -M_{12}^H (Y + X M_{11}) - M_{22}^H M_{21} & qI - M_{12}^H X M_{12} - D^H M_{22} \end{bmatrix} \geq 0 \tag{2.3.81}$$

The representation (2.3.71) is also described as upper Linear Fractional Transformation (LFT, Sinha, 2007) (see Figure 2.3.1).

Yao, Wang, and Li (2009) used this LFT and computed structured singular value as a measure of the bound of frequency response (Sinha, 2007). Again, this measure is in general limited to 2-norm of amplitude vector, rather than $\infty$ norm.

## 2.4    Analytical Computation of the Statistics of Steady-State Amplitudes

From Equations (2.1.1) through (2.1.7) with steady-state $x(t) = \mathbf{a} e^{\iota \omega t}$

$$(K - \omega^2 M + \iota \omega C_t)\mathbf{a} = \mathbf{p}_\ell; \ \iota = \sqrt{-1} \tag{2.4.1}$$

$$\mathbf{a} = [a_1 e^{\iota \psi_1} \quad a_2 e^{\iota \psi_2} \quad a_3 e^{\iota \psi_3} \quad \cdots \quad a_n e^{\iota \psi_n}] \tag{2.4.2}$$

It should be noted that $a_i$ is a real number. Separating real and imaginary parts,

$$H_m \mathbf{u} = \mathbf{f}_{cs} \tag{2.4.3}$$

where

$$\mathbf{u} = [a_1 \cos \psi_1 \quad a_1 \sin \psi_1 \quad a_2 \cos \psi_2 \quad a_2 \sin \psi_2 \quad . \quad . \quad a_1 \cos \psi_n \quad a_1 \sin \psi_n]^T \tag{2.4.4}$$

$$\mathbf{f}_{cs} = f_0 \begin{bmatrix} 1 & 0 & \cos\varphi_2 & \sin\varphi_2 & . & . & \cos\varphi_n & \sin\varphi_n \end{bmatrix}^T \tag{2.4.5}$$

$$H_m = H + \delta H \tag{2.4.6}$$

$$H = \begin{bmatrix} h_1 & h_3 & -h_2 & 0 & 0 & 0 & \cdots & -h_2 & 0 \\ -h_3 & h_1 & 0 & -h_2 & 0 & 0 & \cdots & 0 & -h_2 \\ -h_2 & 0 & h_1 & h_3 & -h_2 & 0 & \cdots & 0 & 0 \\ 0 & -h_2 & -h_3 & h_1 & 0 & -h_2 & \cdots & 0 & 0 \\ \vdots & \vdots & \vdots & \vdots & \vdots & \vdots & \cdots & \vdots & \vdots \\ -h_2 & 0 & 0 & 0 & 0 & 0 & \cdots & h_1 & h_3 \\ 0 & -h_2 & 0 & 0 & 0 & 0 & \cdots & -h_3 & h_1 \end{bmatrix} \tag{2.4.7}$$

$$\delta H = diag \begin{bmatrix} \delta h_{1,1} & \delta h_{1,1} & \delta h_{1,2} & \delta h_{1,2} & \cdots & \delta h_{1,n} & \delta h_{1,n} \end{bmatrix} \tag{2.4.8}$$

$$h_1 = k_t + 2k_c - \omega^2 m_t; \quad h_2 = k_c; \quad h_3 = -c_t\omega \tag{2.4.9}$$

$$\delta h_{1,i} = \delta k_i - \omega^2 \delta m_i; \quad i = 1,2,\ldots,n \tag{2.4.10}$$

Mistuning is modeled by considering perturbations in stiffnesses $\delta k_i$ alone, that is, $\delta m_i = 0$. The amplitude $a_i$ can be obtained as

$$a_i = (u_{2i-1}^2 + u_{2i}^2)^{0.5} = g_i(\delta k_1, \delta k_2, \ldots, \delta k_n) \tag{2.4.11}$$

where $g_i(\ )$ is the function of n random variables.

Given the joint probability distribution of $\delta k_1, \delta k_2, \ldots, \delta k_n$, can we analytically predict the probability distribution function of the amplitude $a_i$? Let the joint probability distribution of $k_1, k_2, \ldots, k_n$ be denoted by $p_k$. For example, if the distribution of each stiffness be independent and Gaussian,

$$p_k = \frac{1}{(\sigma_k \sqrt{2\pi})^n} \prod_{i=1}^{n} \exp(-(k_i - k_t)^2 / 2\sigma_k^2) \tag{2.4.12}$$

where $\sigma_k$ is the standard deviation of each $k_i$. Let (Mignolet and Lin, 1993)

$$\kappa_1 = a_i \tag{2.4.13}$$

$$\kappa_\ell = k_\ell; \quad \ell = 2,3,\ldots,n \tag{2.4.14}$$

Then,

$$p_k = \frac{p_k}{|\det J|} \tag{2.4.15}$$

where the Jacobian matrix $J$ (Papoulis, 1965) is defined as

$$
J = \begin{bmatrix}
\dfrac{\partial a_i}{\partial k_1} & \dfrac{\partial a_i}{\partial k_1} & \cdots & \dfrac{\partial a_i}{\partial k_{n-1}} & \dfrac{\partial a_i}{\partial k_n} \\
0 & 1 & \cdots & 0 & 0 \\
\vdots & \vdots & \cdots & \vdots & \vdots \\
0 & 0 & \cdots & 1 & 0 \\
0 & 0 & \cdots & 0 & 1
\end{bmatrix}
\tag{2.4.16}
$$

Therefore,

$$
\det J = \frac{\partial a_i}{\partial k_1} = \frac{1}{2a_i}\frac{\partial a_i^2}{\partial k_1}
\tag{2.4.17}
$$

From Equations (2.4.15) and (2.4.17),

$$
p_k = \frac{p_{\mathbf{k}}}{|\det J|} = 2a_i \frac{p_{\mathbf{k}}}{\left|\dfrac{\partial a_i^2}{\partial k_1}\right|}
\tag{2.4.18}
$$

The partial derivative of $a_i^2$ with respect to $k_i$ is derived as follows. From Equation (2.4.3),

$$
u_{2i-1} = \mathbf{e}_{2i-1}^T \mathbf{u} = \mathbf{e}_{2i-1}^T H_m^{-1}\mathbf{f}_{cs} \quad \text{and} \quad u_{2i} = \mathbf{e}_{2i}^T \mathbf{u} = \mathbf{e}_{2i}^T H_m^{-1}\mathbf{f}_{cs}
\tag{2.4.19}
$$

where

$$
\mathbf{e}_{2i-1}^T = \begin{bmatrix} 0 & 0 & \cdots & 1 & 0 & \cdots & 0 & 0 \end{bmatrix}
\tag{2.4.20}
$$

$$
\mathbf{e}_{2i}^T = \begin{bmatrix} 0 & 0 & \cdots & 0 & 1 & \cdots & 0 & 0 \end{bmatrix}
\tag{2.4.21}
$$

It should be noted that the $\ell th$ element of $\mathbf{e}_\ell^T$ is one, and all other elements are zeros. Next,

$$
a_i^2 = u_{2i-1}^2 + u_{2i}^2 = u_{2i-1}u_{2i-1}^T + u_{2i}u_{2i}^T
\tag{2.4.22}
$$

Substituting Equations (2.4.19) and (2.4.20) into Equation (2.4.22),

$$
a_i^2 = \mathbf{e}_{2i-1}^T H_m^{-1}\mathbf{f}_{cs}\mathbf{f}_{cs}^T H_m^{-T}\mathbf{e}_{2i-1} + \mathbf{e}_{2i}^T H_m^{-1}\mathbf{f}_{cs}\mathbf{f}_{cs}^T H_m^{-T}\mathbf{e}_{2i}
\tag{2.4.23}
$$

It can be easily shown that

$$
\frac{\partial H_m^{-1}}{\partial k_1} = -H_m^{-1}\frac{\partial H_m}{\partial k_1}H_m^{-1}
\tag{2.4.24}
$$

Utilizing Equation (2.4.24),

$$\frac{\partial a_i^2}{\partial k_1} = -\mathbf{e}_{2i-1}^T H_m^{-1} \frac{dH_m}{dk_1} H_m^{-1}\mathbf{f}_{cs}\mathbf{f}_{cs}^T H_m^{-T}\mathbf{e}_{2i-1} - \mathbf{e}_{2i-1}^T H_m^{-1}\mathbf{f}_{cs}\mathbf{f}_{cs}^T H_m^{-T}\frac{dH_m^T}{dk_1} H_m^{-T}\mathbf{e}_{2i-1}$$

$$- \mathbf{e}_{2i}^T H_m^{-1}\frac{dH_m}{dk_1} H_m^{-1}\mathbf{f}_{cs}\mathbf{f}_{cs}^T H_m^{-T}\mathbf{e}_{2i} - \mathbf{e}_{2i-1}^T H_m^{-1}\mathbf{f}_{cs}\mathbf{f}_{cs}^T H_m^{-T}\frac{dH_m^T}{dk_1} H_m^{-T}\mathbf{e}_{2i}$$

$$(2.4.25)$$

Because the transpose of any scalar equals that scalar,

$$\mathbf{e}_{2i-1}^T H_m^{-1}\frac{dH_m}{dk_1} H_m^{-1}\mathbf{f}_{cs}\mathbf{f}_{cs}^T H_m^{-T}\mathbf{e}_{2i-1} = \mathbf{e}_{2i-1}^T H_m^{-1}\mathbf{f}_{cs}\mathbf{f}_{cs}^T H_m^{-T}\frac{dH_m^T}{dk_1} H_m^{-T}\mathbf{e}_{2i-1} \quad (2.4.26)$$

$$\mathbf{e}_{2i}^T H_m^{-1}\frac{dH_m}{dk_1} H_m^{-1}\mathbf{f}_{cs}\mathbf{f}_{cs}^T H_m^{-T}\mathbf{e}_{2i} = \mathbf{e}_{2i-1}^T H_m^{-1}\mathbf{f}_{cs}\mathbf{f}_{cs}^T H_m^{-T}\frac{dH_m^T}{dk_1} H_m^{-T}\mathbf{e}_{2i} \quad (2.4.27)$$

Using Equations (2.4.26) and (2.4.27),

$$\frac{\partial a_i^2}{\partial k_1} = -2\mathbf{e}_{2i-1}^T H_m^{-1}\frac{\partial H_m}{\partial k_1} H_m^{-1}\mathbf{f}_{cs}\mathbf{f}_{cs}^T H_m^{-T}\mathbf{e}_{2i-1} - 2\mathbf{e}_{2i}^T H_m^{-1}\frac{\partial H_m}{\partial k_1} H_m^{-1}\mathbf{f}_{cs}\mathbf{f}_{cs}^T H_m^{-T}\mathbf{e}_{2i}$$

$$(2.4.28)$$

From Equation (2.4.18), the probability density function of any amplitude $a_i$ is given by

$$p_{a_i} = \int_{-\infty}^{\infty} \cdots \int_{-\infty}^{\infty} 2a_i \frac{p_\mathbf{k}}{\left|\dfrac{\partial a_i^2}{\partial k_1}\right|} dk_2 \cdots dk_n \qquad (2.4.29)$$

The relationship is exact, but unfortunately involves multidimensional integration, which cannot be easily computed. Therefore, computationally efficient approximate methods are sought.

## 2.4.1    Neumann Expansion

From Equations (2.4.3) and (2.4.6),

$$\mathbf{u} = (H + \delta H)^{-1}\mathbf{f}_{cs} = (I + H^{-1}\delta H)^{-1}\mathbf{u}_t \qquad (2.4.30)$$

where

$$\mathbf{u}_t = H^{-1}\mathbf{f}_{cs} \qquad (2.4.31)$$

Assuming that $\|H^{-1}\delta H\|_2 < 1$,

$$(I + H^{-1}\delta H)^{-1} = I - H^{-1}\delta H + (H^{-1}\delta H)^2 - (H^{-1}\delta H)^3 + \cdots \qquad (2.4.32)$$

At resonant condition, $\|H^{-1}\|_2$ is large, particularly with small damping. Hence, the condition $\|H^{-1}\delta H\|_2 < 1$ is satisfied for small values of $\delta k_i$ for small damping and resonant condition.

### First Order Approximation

Neglecting higher-order terms (Sinha, 1986),

$$(I + H^{-1}\delta H)^{-1} \approx I - H^{-1}\delta H \tag{2.4.33}$$

Substituting Equation (2.4.33) into Equation (2.4.30),

$$\mathbf{u} \approx \mathbf{u}_t - H^{-1}\delta H \mathbf{u}_t \tag{2.4.34}$$

The pth element of the vector $\mathbf{u}$ is written as

$$u_p = u_{tp} - \sum_{\ell=1}^{n} (hi_{p,2\ell-1} u_{t,2\ell-1} + hi_{p,2\ell} u_{t,2\ell})\delta h_{1,\ell}; \quad p = 1,2,\ldots,2n \tag{2.4.35}$$

where

$$(H^{-1})_{p\ell} = hi_{p\ell} \tag{2.4.36}$$

and $\delta h_{1,\ell}$ is defined by Equation (2.4.10). It is assumed that the random variables $\delta h_{1,\ell}$ are Gaussian and independent with zero mean. From Equation (2.4.35),

$$\bar{u}_p = E(u_p) = u_{tp} - \sum_{\ell=1}^{n} (hi_{p,2\ell-1} u_{t,2\ell-1} + hi_{p,2\ell} u_{t,2\ell})E(\delta h_{1,\ell}) = u_{tp} \tag{2.4.37}$$

And,

$$\sigma_p^2 = E((u_p - u_{tp})^2) = \sum_{\ell=1}^{n} (hi_{p,2\ell-1} u_{t,2\ell-1} + hi_{p,2\ell} u_{t,2\ell})^2 \, E(\delta h_{1,\ell}^2) \tag{2.4.38}$$

In this case, it is easy to see from Equation (2.4.35) that $u_p$ will be Gaussian with mean $= u_{tp}$ and the variance given by Equation (2.4.38). From Equation (2.4.37), the correlation coefficient between $u_{2j-1}$ and $u_{2j}$ is obtained as

$$\gamma_j = \frac{E((u_{2j-1} - \bar{u}_{2j-1})(u_{2j} - \bar{u}_{2j}))}{\sigma_{2j-1}\sigma_{2j}} \tag{2.4.39}$$

Note that the Equation (2.4.35) also yields

$$E((u_p - u_{tp})(u_q - u_{tq})) = \sum_{\ell=1}^{n} (hi_{p,2\ell-1} u_{t,2\ell-1} + hi_{p,2\ell} u_{t,2\ell}) \tag{2.4.40}$$
$$(hi_{q,2\ell-1} u_{t,2\ell-1} + hi_{q,2\ell} u_{t,2\ell})E(\delta h_{1,\ell}^2)$$

It can also be shown that any linear combination of $u_{2j-1}$ and $u_{2j}$ will also be Gaussian. Therefore, $u_{2j-1}$ and $u_{2j}$ will be jointly Gaussian (Papoulis, 1965), that is,

$$f(u_{2j-1}, u_{2j}) = \frac{1}{2\pi\sigma_{2j-1}\sigma_{2j}(1-\gamma_j^2)^{0.5}}$$
$$\times \exp\left[\frac{(u_{2j-1}-\bar{u}_{2j-1})^2}{\sigma_{2j-1}^2} - \frac{2\gamma_j(u_{2j-1}-\bar{u}_{2j-1})(u_{2j}-\bar{u}_{2j})}{\sigma_{2j-1}\sigma_{2j}} + \frac{(u_{2j}-\bar{u}_{2j})^2}{\sigma_{2j}^2}\right] \quad (2.4.41)$$

From the relationship (2.4.11), the probability distribution function of the amplitude $a_j$ is given by

$$g(a_j) = \frac{a_j}{2\pi\sigma_{2j-1}\sigma_{2j}(1-\gamma_j^2)^{0.5}} \int_0^{2\pi} \exp\left[-\frac{z(\psi_j)}{2(1-\gamma_j^2)}\right] d\psi_j \quad (2.4.42)$$

where

$$z(\psi_j) = \frac{(a_j\cos\psi_j-\bar{u}_{2j-1})^2}{\sigma_{2j-1}^2} - \frac{2\gamma_j(a_j\cos\psi_j-\bar{u}_{2j-1})(a_j\sin\psi_j-\bar{u}_{2j})}{\sigma_{2j-1}\sigma_{2j}}$$
$$+ \frac{(a_j\sin\psi_j-\bar{u}_{2j})^2}{\sigma_{2j}^2} \quad (2.4.43)$$

Let

$$\delta u_{2j-1} = u_{2j-1} - u_{t,2j-1} = \delta a_j \cos\delta\psi_j \quad (2.4.44)$$

$$\delta u_{2j} = u_{2j} - u_{t,2j} = \delta a_j \sin\delta\psi_j \quad (2.4.45)$$

Then,

$$\delta a_j = \sqrt{\delta u_{2j-1}^2 + \delta u_{2j}^2} \quad (2.4.46)$$

A closed-form expression for the probability distribution of deviation in the amplitude $\delta a_j$ is obtained (Sinha, 1986)

$$g(\delta a_j) = \frac{\delta a_j}{\sigma_{2j-1}\sigma_{2j}(1-\gamma_j^2)^{0.5}} \exp\left[-\frac{(\delta a_j)^2}{4(1-\gamma_j^2)}\left(\frac{1}{\sigma_{2j-1}^2}+\frac{1}{\sigma_{2j}^2}\right)\right] I_{b1}(v) \quad (2.4.47)$$

where

$$v = \frac{(\delta a_j)^2}{2(1-\gamma_j^2)}\left[\frac{1}{4}\left(\frac{1}{\sigma_{2j-1}^2}-\frac{1}{\sigma_{2j}^2}\right)^2 + \left(\frac{\gamma_j}{\sigma_{2j-1}\sigma_{2j}}\right)^2\right]^{0.5} \quad (2.4.48)$$

and $I_{b1}$ is Bessel function of first kind.

### Higher-Order Approximation

Neglecting fourth and higher-order terms in Equation (2.4.32) (Sinha and Chen, 1989),

$$(I + H^{-1}\delta H)^{-1} \approx I - H^{-1}\delta H + (H^{-1}\delta H)^2 - (H^{-1}\delta H)^3 \qquad (2.4.49)$$

Substituting Equation (2.4.49) into Equation (2.4.30),

$$\mathbf{u} \approx \mathbf{u}_t - H^{-1}\delta H \mathbf{u}_t + +(H^{-1}\delta H)^2 \mathbf{u}_t - (H^{-1}\delta H)^3 \mathbf{u}_t \qquad (2.4.50)$$

$$u_p = u_{tp} - \sum_{\ell=1}^{n} w1_{p,\ell} \delta h_{1,\ell} + \sum_{\ell=1}^{n} \sum_{i=1}^{n} w2_{p,\ell,i} \delta h_{1,\ell} \delta h_{1,i}$$

$$+ \sum_{\ell=1}^{n} \sum_{i=1}^{n} \sum_{\eta=1}^{n} w3_{p,\ell,i,\eta} \delta h_{1,\ell} \delta h_{1,i} \delta h_{1,\eta} \qquad (2.4.51)$$

where

$$w1_{p,\ell} = hi_{p,2\ell-1} u_{t,2\ell-1} + hi_{p,2\ell} u_{t,2\ell} \qquad (2.4.52)$$

$$w2_{p,\ell,i} = hi_{p,2\ell-1} w1_{2\ell-1,i} + hi_{p,2\ell} w1_{2\ell,i} \qquad (2.4.53)$$

$$w3_{p,\ell,i,\eta} = hi_{p,2\ell-1} w2_{2\ell-1,i,\eta} + hi_{p,2\ell} w1_{2\ell,i,\eta} \qquad (2.4.54)$$

Using central limit theorem, Sinha and Chen (1989) has shown that the distribution of $u_p$ can be assumed to be Gaussian. Further, $u_{2j-1}$ and $u_{2j}$ are also approximated as jointly Gaussian.

For Gaussian $\delta h_{1,\ell} = \delta k_\ell$ (Papoulis, 1965),

$$E(\delta h_{1,\ell}^v) = \begin{cases} 0 & \textit{if } v \textit{ odd} \\ \dfrac{\sigma^v v!}{2^{v/2}(v/2)!} & \textit{if } v \textit{ even} \end{cases} \qquad (2.4.55)$$

Further, because $\delta h_{1,\ell}$ and $\delta h_{1,i}$, $\ell \neq i$, are assumed to be independent random variables,

$$E[\delta h_{1,\ell}^v \delta h_{1,i}^\varsigma] = E[\delta h_{1,\ell}^v]E[\delta h_{1,i}^\varsigma]; \ \ell \neq i; \text{ integers } v \text{ and } \varsigma \qquad (2.4.56)$$

From Equation (2.4.51),

$$\bar{u}_p = E(u_p) = u_{tp} - \sigma^2 \sum_{\ell=1}^{n} w2_{p,\ell,\ell} \qquad (2.4.57)$$

$$E((u_p - u_{tp})(u_q - u_{tq})) = term1 + term2 + term3 + term4 \qquad (2.4.58)$$

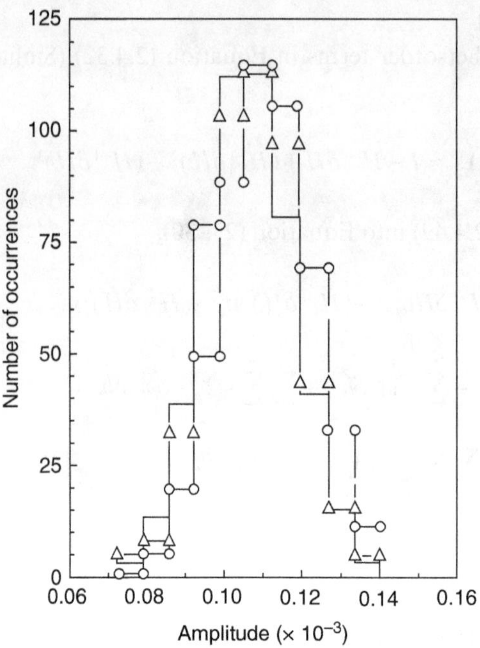

**Figure 2.4.1.**    Distribution of amplitude, std. dev. $\sigma_k = 5000\,N\,/\,m$, o: linear term, $\Delta$: higher order, ____ Monte Carlo (Sinha and Chen, 1989).

where

$$term1 = \left[\sum_{\ell=1}^{1} w1_{p,\ell}\,w1_{q,\ell}\right]\sigma^2 \tag{2.4.59}$$

$$term2 = \left[3\sum_{\ell=1}^{n} w3_{q,\ell,\ell,\ell}\,w1_{p,\ell} + \sum_{\ell=1}^{n}\sum_{\substack{i=1\\ \ell\neq i}}^{n} w1_{p,i}\,(w3_{q,\ell,\ell,i} + w3_{q,\ell,i,\ell}\,w3_{q,i,\ell,\ell})\right]\sigma^4 \tag{2.4.60}$$

$$term3 = \left[3\sum_{\ell=1}^{n} w3_{p,\ell,\ell,\ell}\,w1_{q,\ell} + \sum_{\ell=1}^{n}\sum_{\substack{i=1\\ \ell\neq i}}^{n} w1_{q,i}\,(w3_{p,\ell,\ell,i} + w3_{p,\ell,i,\ell}\,w3_{p,i,\ell,\ell})\right]\sigma^4 \tag{2.4.61}$$

$$term4 = \left[3\sum_{\ell=1}^{n} w2_{p,\ell,\ell}\,w2_{q,\ell,\ell} + \sum_{\ell=1}^{n}\sum_{\substack{i=1\\ \ell\neq i}}^{n} w2_{p,\ell,\ell}\,w2_{q,i,i} + w2_{p,\ell,i}\,(w2_{q,\ell,i} + w2_{q,i,\ell})\right]\sigma^4 \tag{2.4.62}$$

Distributions of amplitudes predicted by Equation (2.4.42) with Equations (2.4.57) and (2.4.58) for mean and standard deviations are compared to those from Monte Carlo simulations; see Figures 2.4.1 to 2.4.3 for standard deviations of

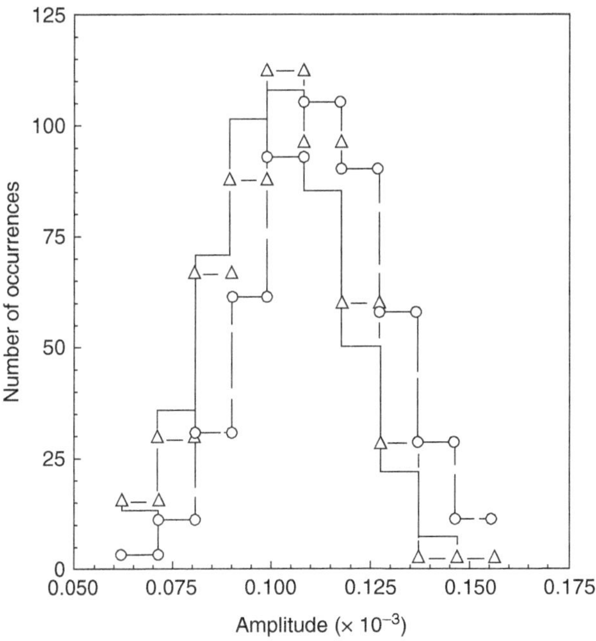


**Figure 2.4.2.** Distributions of amplitude, std. dev. $\sigma_k = 7500\,N/m$ o: linear term, $\Delta$: higher order, _____ Monte Carlo (Sinha and Chen, 1989).

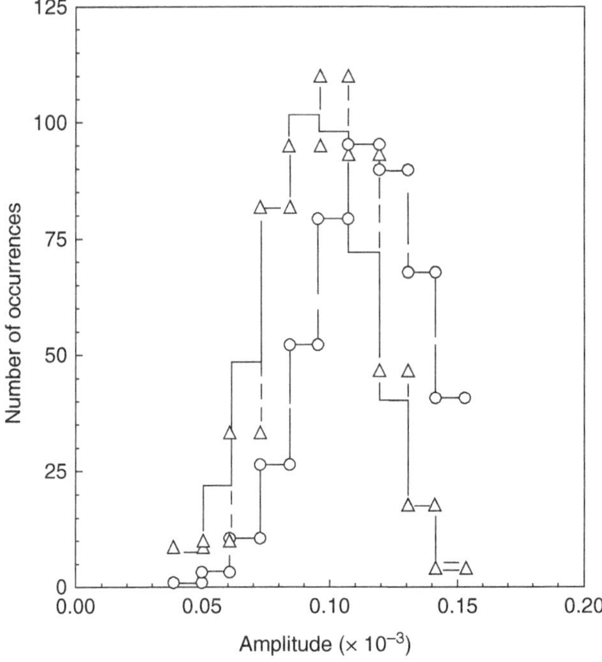


**Figure 2.4.3.** Distribution of amplitude, std. dev. $\sigma_k = 10000\,N/m$ o: linear term, $\Delta$: higher order, _____ Monte Carlo (Sinha and Chen, 1989).

stiffness $\sigma = \sigma_k$ = 5,000 N/m, 7,500 N/m and 10,000 N/m. System parameters (SI Units) are $k_t$ = 430000, $k_c$ = 45430, and $c_t$ = 1.38. In Figure 2.4.1, results for linear approximations are also shown.

### 2.4.2.        Polynomial Chaos

Neumann expansions have guaranteed stochastic convergence only when the standard deviation of perturbations due to mistuning is less than a certain maximum value at a damping level. Therefore, Sinha (2006c) has used the method of polynomial chaos, originally developed by Weiner (1958) and used by Ghanem and Spanos (1990) to develop stochastic finite elements. The advantage of this method is that any non-Gaussian distribution can be expanded in terms of polynomial chaoses with a guaranteed convergence, provided a sufficient number of polynomial chaoses is selected. From Equation (1.1.9),

$$\Delta K = \sum_{\ell=1}^{n} \xi_\ell K_f^\ell; \tag{2.4.63}$$

$$K_f^i = \sigma \begin{bmatrix} 0 & \cdots & 0 & 0 & 0 & \cdots & 0 \\ \cdots & \cdots & \cdots & \cdots & \cdots & \cdots & \cdots \\ 0 & \cdots & 0 & 0 & 0 & \cdots & 0 \\ 0 & \cdots & 0 & 1 & 0 & \cdots & 0 \\ 0 & \cdots & 0 & 0 & 0 & \cdots & 0 \\ \cdots & \cdots & \cdots & \cdots & \cdots & \cdots & \cdots \\ 0 & \cdots & 0 & 0 & 0 & \cdots & 0 \end{bmatrix} \tag{2.4.64}$$

where $\sigma$ is the standard deviation of each random variable $\delta k_\ell$. Further, $\xi_\ell$ is a random variable with zero mean and unity standard deviation. From Equations (2.4.1) and (2.4.63),

$$\left[ I + \sum_{\ell=1}^{n} \xi_\ell Q^\ell(\omega) \right] \mathbf{a} = \mathbf{g} \tag{2.4.65}$$

where

$$\mathbf{g} = (K_t - \omega^2 M_t + j\omega C_t)^{-1} \mathbf{p}_\ell \tag{2.4.66}$$

$$Q^\ell(\omega) = (K_t - \omega^2 M_t + j\omega C_t)^{-1} K_f^\ell \tag{2.4.67}$$

Using polynomial chaos, the solution of Equation (2.4.65) can be written as

$$\mathbf{a} = a_{i0}\Gamma_0 + \sum_{i_1=1}^{n} a_{i_1}\Gamma_1(\xi_{i_1}) + \sum_{i_1=1}^{n}\sum_{i_2=1}^{i_1} a_{i_1,i_2}\Gamma_2(\xi_{i_1},\xi_{i_2})$$

$$+ \sum_{i_1=1}^{n}\sum_{i_2=1}^{i_1}\sum_{i_3=1}^{i_2} a_{i_1,i_2,i_3}\Gamma_3(\xi_{i_1},\xi_{i_2},\xi_{i_3}) + \cdots \tag{2.4.68}$$

where

$$\Gamma_0 = 1 \quad \Gamma_1(\xi_{i_1}) = \xi_{i_1} \quad \Gamma_2(\xi_{i_1}, \xi_{i_2}) = \xi_{i_1}\xi_{i_2} - \delta_{i_1, i_2}$$

$$\Gamma_3(\xi_{i_1}, \xi_{i_2}, \xi_{i_3}) = \xi_{i_1}\xi_{i_2}\xi_{i_3} - \xi_{i_1}\delta_{i_2 i_3} - \xi_{i_2}\delta_{i_1 i_3} - \xi_{i_3}\delta_{i_1 i_2} \tag{2.4.69}$$

and $\delta_{ij}$ is the Kronecker delta. Alternatively,

$$\mathbf{a} = \sum_{\ell=0}^{p} \beta_\ell \Psi_\ell(\boldsymbol{\xi}) \tag{2.4.70}$$

$$\Psi_0(\boldsymbol{\xi}) = \Gamma_0; \Psi_1(\boldsymbol{\xi}) = \Gamma_1(\xi_1), \ldots, \Psi_{n+1}(\boldsymbol{\xi}) = \Gamma_1(\xi_n), \Psi_{n+2}(\boldsymbol{\xi}) = \Gamma_2(\xi_1, \xi_1), \cdots \tag{2.4.71}$$

It should be noted that

$$E[\Psi_\ell(\boldsymbol{\xi})] = 0; \quad \ell > 0 \tag{2.4.72}$$

$$E[\Psi_\ell(\boldsymbol{\xi})\Psi_i(\boldsymbol{\xi})] = 0; \quad \ell \neq i \tag{2.4.73}$$

Substituting Equation (2.4.70) into Equation (2.4.65),

$$\left[ I + \sum_{\ell=1}^{n} \xi_\ell Q^\ell(\omega) \right] \sum_{\ell=0}^{p} \beta_\ell \Psi_\ell(\boldsymbol{\xi}) = \mathbf{g} \tag{2.4.74}$$

Multiplying Equation (2.4.74) by $\Psi_i(\boldsymbol{\xi})$ and taking expected values on both sides

$$\beta_i \rho_i I + \sum_{v=1}^{p} \sum_{\ell=1}^{n} \chi(\ell, v, i) Q^\ell \beta_v = E[g\Psi_i(\boldsymbol{\xi})]; \ i = 0, 1, 2, \ldots, p \tag{2.4.75}$$

where

$$\rho_i = E(\Psi_i^2) \quad \text{and} \quad \chi(\ell, v, i) = E(\xi_\ell \Psi_v \Psi_i) \tag{2.4.76}$$

The coefficients $\rho_i$ and $\chi(\ell, v, i)$ have been calculated analytically by Sinha (2006c). Coefficients $\beta_i$, $i = 0, 1, 2, \ldots, p$ are found by solving a linear system of equations.

Distributions of amplitudes and maximum amplitudes (among all blades) obtained by Equations (2.4.70) are compared to those from Monte Carlo simulations, see Figures 2.4.4–2.4.7 for standard deviations $\sigma = \sigma_k = 1,000, 4,000, 7,000$, and $10,000$ N/m. System parameters (SI Units) are $k_t = 430000$, $k_c = 45430$, and $c_t = 1.38$.

## 2.4.3 An Open Mathematical Problem

Consider

$$(H + \delta H)\mathbf{a} = \mathbf{f} \tag{2.4.77}$$

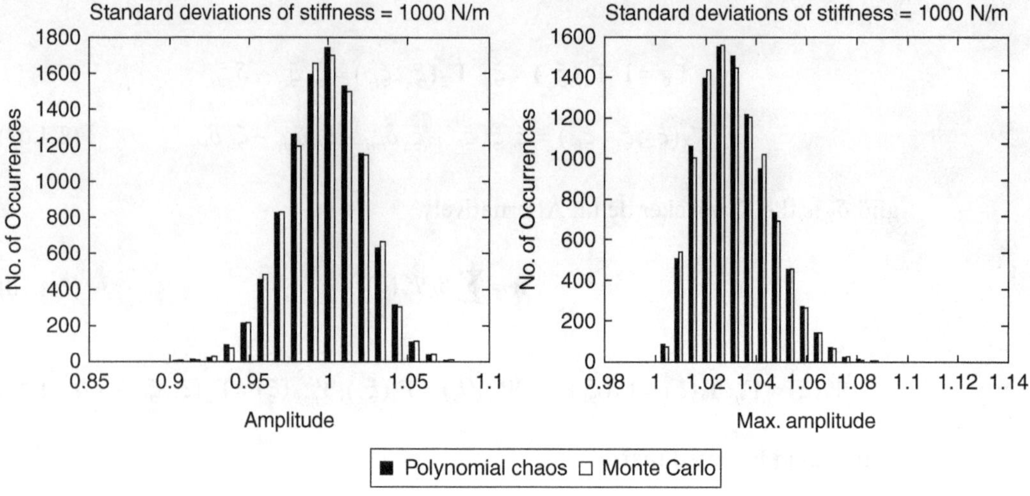

**Figure 2.4.4.**   Distributions of amplitude and maximum amplitude (std. dev. $\sigma_k = 1000\,N\,/\,m$) (Sinha, 2006c).

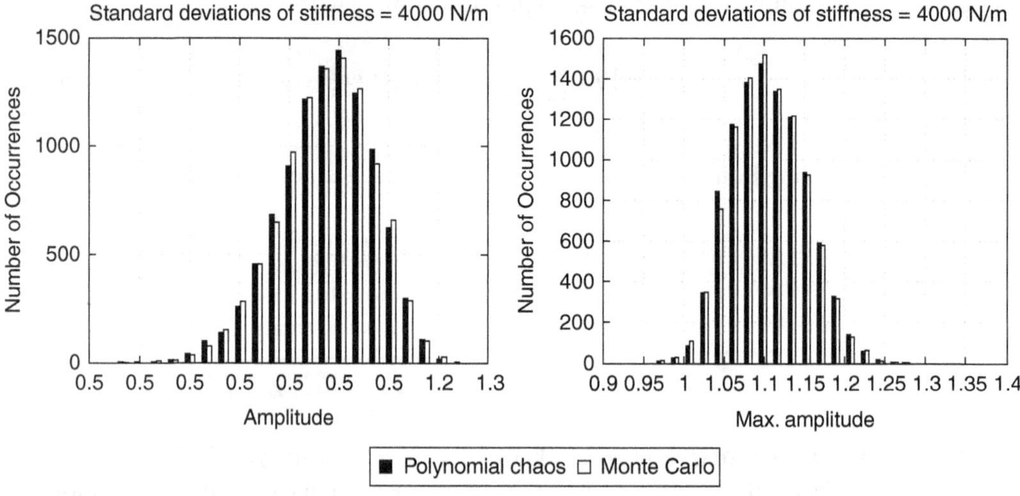

**Figure 2.4.5.**   Distributions of amplitude and maximum amplitude (std. dev. $\sigma_k = 4000\,N\,/\,m$) (Sinha, 2006c).

The matrices $H$ and $\delta H$ are Hermitian. The matrix $H$ is circulant, deterministic, and *near singular* and is a function of the frequency $\omega$. The vector $\mathbf{f}$ is deterministic. The matrix $\delta H$ is block-diagonal and random with a finite number of independent random variables.

At present, the analytical computation of the statistics of the peak maximum amplitude $\max\limits_{\omega} \|\mathbf{a}\|_\infty$ is not possible (Tao, 2013).

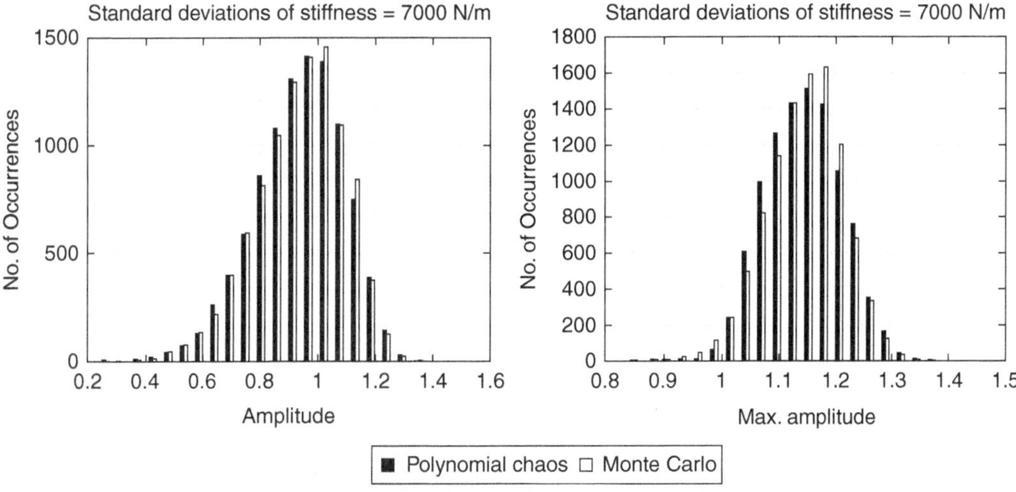

**Figure 2.4.6.**  Distributions of amplitude and maximum amplitude (std. dev. $\sigma_k = 7000\,N\,/\,m$) (Sinha, 2006c).

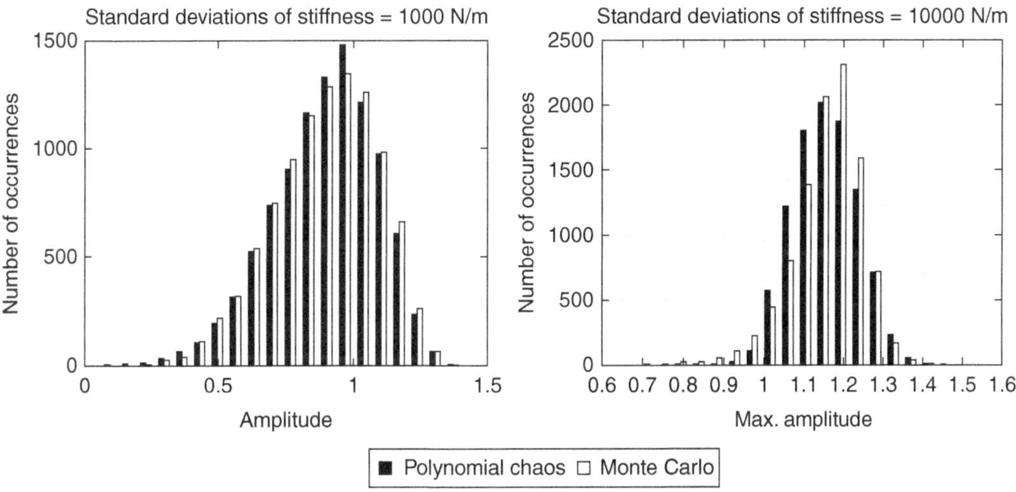

**Figure 2.4.7.**  Distributions of amplitude and maximum amplitude (std. dev. $\sigma_k = 10000\,N\,/\,m$)(Sinha, 2006c).

# 3 Reduced-Order Models and Response of Nearly Periodic Structures

First, equivalent sector and unit cell analyses are presented for structures having periodicity in one and two dimensions, respectively. Next, reduced-order models of a mistuned bladed disk based on a subset of tuned bladed-disk modes, single-family tuned modes, and component mode synthesis approach are presented. Then, the MMDA is presented to obtain a high-fidelity reduced-order model of a bladed rotor with geometric mistuning along with numerical results for academic and industrial-scale integrally bladed rotors. Next, MMDA approach is extended to identify mistuning parameters from measured mistuned modes and frequencies. Lastly, a MMDA-based reduced-order model algorithm is presented for a multi-stage bladed rotor with geometric mistuning.

## 3.1 Analysis of Periodic Structures

### 3.1.1 Derivation of the Equivalent Mass and Stiffness Matrices Directly from a Sector of a Bladed Disk

Describe the tuned mass and stiffness matrices of the full (360 degrees) rotor as follows:

$$
M_t = \begin{bmatrix}
M_{t,1,1} & M_{t,1,2} & \cdots & M_{t,1,n} \\
M_{t,2,1} & M_{t,2,2} & \cdots & M_{t,2,n} \\
\vdots & \vdots & \vdots & \vdots \\
M_{t,n,1} & M_{t,n,2} & \cdots & M_{t},n,n
\end{bmatrix}
\tag{3.1.1}
$$

$$
K_t = \begin{bmatrix}
K_{t,1,1} & K_{t,1,2} & \cdots & K_{t,1,n} \\
K_{t,2,1} & K_{t,2,2} & \cdots & K_{t,2,n} \\
\vdots & \vdots & \vdots & \vdots \\
K_{t,n,1} & K_{t,n,2} & \cdots & K_{t},n,n
\end{bmatrix}
\tag{3.1.2}
$$

Matrices $M_t$ and $K_t$ are block circulant, that is, the second block row will be obtained by shifting the first block row by one location, the third block row will be obtained by shifting the second block row by one location, and so on. Further, the matrices $M_t$ and $K_t$ are symmetric. Therefore, certain relationships have to be satisfied, for example,

$$M_{t,1,n} = M_{t,1,2}^T; \; K_{t,1,n} = K_{t,1,2}^T; \dots \dots \quad (3.1.3)$$

For a rotationally periodic structure with three sectors, mass and stiffness matrices will be as follows:

$$M_t = \begin{bmatrix} M_{t,1,1} & M_{t,1,2} & M_{t,1,2}^T \\ M_{t,1,2}^T & M_{t,1,1} & M_{t,1,2} \\ M_{t,1,2} & M_{t,1,2}^T & M_{t,1,1} \end{bmatrix} \quad \text{and} \quad K_t = \begin{bmatrix} K_{t,1,1} & K_{t,1,2} & K_{t,1,2}^T \\ K_{t,1,2}^T & K_{t,1,1} & K_{t,1,2} \\ K_{t,1,2} & K_{t,1,2}^T & K_{t,1,1} \end{bmatrix} \quad (3.1.4)$$

Let the differential equations of motion of the full (360 degrees) rotor be

$$M_t \ddot{\mathbf{u}} + K_t \mathbf{u} = 0 \quad (3.1.5)$$

where

$$\mathbf{u} = \begin{bmatrix} \mathbf{u}_1 \\ \mathbf{u}_2 \\ \vdots \\ \mathbf{u}_{n-1} \\ \mathbf{u}_n \end{bmatrix} \quad (3.1.6)$$

and $\mathbf{u}_i$ is the displacement vector associated with the sector number $i$.

Mode shapes are cyclic with $n$ distinct values of interblade/intersector phase angles:

$$\psi_i = \frac{2\pi i}{n}; \; i = 0, 1, \dots, n-1 \quad (3.1.7)$$

Then, the nodal displacement vector of sector $\ell$ is related to that of sector 1 as follows:

$$\mathbf{u}_\ell = \mathbf{u}_1 e^{j(\ell-1)\psi_i} : \quad \ell = 1, 2, \dots, n; \; j = \sqrt{-1} \quad (3.1.8)$$

Substituting Equations (3.1.6) and (3.1.8) into Equation (3.1.5), equivalent single-sector dynamics is represented as

$$M_t^c \ddot{\mathbf{u}}_1 + K_t^c \mathbf{u}_1 = 0 \quad (3.1.9)$$

where mass and stiffness matrices of the equivalent single sector are complex, and described as follows:

$$M_t^c = M_{t,1,1} + M_{t,1,2} e^{j\psi_i} + M_{t,1,3} e^{j2\psi_i} + \cdots + M_{t,1,n} e^{j(n-1)\psi_i} \quad (3.1.10)$$

$$K_t^c = K_{t,1,1} + K_{t,1,2}e^{j\psi_i} + K_{t,1,3}e^{j2\psi_i} + \cdots + K_{t,1,n}e^{j(n-1)\psi_i} \qquad (3.1.11)$$

For a typical structure,

$$M_{t,1,3} = 0, \ldots, M_{t,1,n-1} = 0 \qquad (3.1.12)$$

$$K_{t,1,3} = 0, \ldots, K_{t,1,n-1} = 0 \qquad (3.1.13)$$

In this case,

$$M_t^c = M_{t,1,1} + M_{t,1,2}e^{j\psi_i} + M_{t,1,n}e^{-j\psi_i} \qquad (3.1.14)$$

$$K_t^c = K_{t,1,1} + K_{t,1,2}e^{j\psi_i} + K_{t,1,n}e^{-j\psi_i} \qquad (3.1.15)$$

For a simple spring-mass model (Figure 1.1.1), $M_{t,1,2} = 0$ *and* $M_{t,1,n-1} = 0$; $K_{t,1,2} = K_{t,1,n-1}$. In such cases, both $M_t^c$ and $K_t^c$ are real matrices. But, in general, for matrices $K_t$ and $M_t$ to be symmetric, condition (3.1.3) is satisfied, and both $M_t^c$ and $K_t^c$ are complex matrices. Separating real and imaginary parts,

$$M_t^c = M_{t,r\ell}^c + jM_{t,im}^c \quad \text{and} \quad K_t^c = K_{t,r\ell}^c + jK_{t,im}^c; \; j = \sqrt{-1} \qquad (3.1.16a, b)$$

From Equations (3.1.14) and (3.1.15),

$$M_{t,r\ell}^c = M_{t,1,1} + (M_{t,1,2} + M_{t,1,2}^T)\cos\psi_i; \quad M_{t,im}^I = (M_{t,1,2} - M_{t,1,2}^T)\sin\psi_i \qquad (3.1.17a, b)$$

$$K_{t,r\ell}^c = K_{t,1,1} + (K_{t,1,2} + K_{t,1,2}^T)\cos\psi_i; \quad K_{t,im}^I = (K_{t,1,2} - K_{t,1,2}^T)\sin\psi_i \qquad (3.1.18a, b)$$

It can be seen that real parts of complex matrices, $M_{t,r\ell}^c$ and $K_{t,r\ell}^c$, are symmetric, whereas imaginary parts, $M_{t,im}^c$ and $K_{t,im}^c$, are skew-symmetric. Let

$$\mathbf{u}_1 = \mathbf{u}_{r\ell} + j\mathbf{u}_{im} \qquad (3.1.19)$$

Substituting Equations (3.1.14)–(3.1.19) into Equation (3.1.9), and equating real and imaginary parts,

$$\begin{bmatrix} M_{t,r\ell}^c & -M_{t,im}^c \\ M_{t,im}^c & M_{t,r\ell}^c \end{bmatrix}\begin{bmatrix} \ddot{\mathbf{u}}_{r\ell} \\ \ddot{\mathbf{u}}_{im} \end{bmatrix} + \begin{bmatrix} K_{t,r\ell}^c & -K_{t,im}^c \\ K_{t,im}^c & K_{t,r\ell}^c \end{bmatrix}\begin{bmatrix} \mathbf{u}_{r\ell} \\ \mathbf{u}_{im} \end{bmatrix} = \begin{bmatrix} \mathbf{0} \\ \mathbf{0} \end{bmatrix} \qquad (3.1.20)$$

## Derivation of the Equivalent Mass and Stiffness Matrices Directly from a Sector

Let the displacement vector of the sector $s$ (Figure 3.1.1) be denoted as

$$\mathbf{u}_s = \begin{bmatrix} \mathbf{u}_L \\ \mathbf{u}_b \\ \mathbf{u}_R \end{bmatrix} \qquad (3.1.21)$$

where $\mathbf{u}_L$, $\mathbf{u}_b$ and $\mathbf{u}_R$ are displacements of nodes on left boundary, interior nodes, and nodes on the right boundary, respectively. Let $\mathbf{f}_{s-1}$ be the force on the boundary

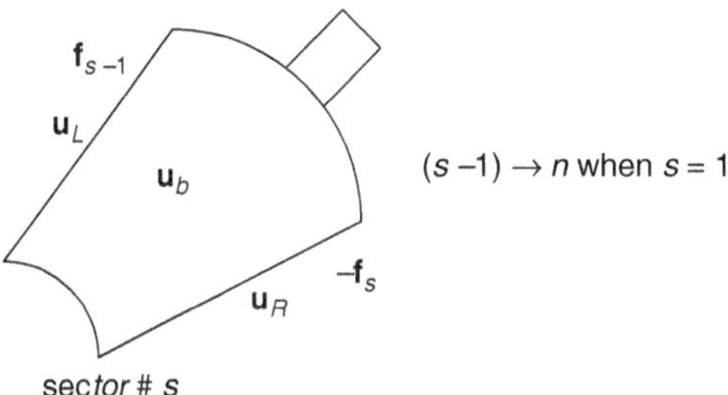


**Figure 3.1.1.** A sector of a periodic structure.

nodes from sector number $s-1$ to sector number $s$. Assuming that interior nodes have zero forces, equations of motion of sector $s$ can be written as

$$M_s \ddot{\mathbf{u}}_s + K_s \mathbf{u}_s = \mathbf{fv}_s \tag{3.1.22}$$

where

$$\mathbf{fv}_s = \begin{bmatrix} \mathbf{f}_{s-1} \\ \mathbf{0} \\ -\mathbf{f}_s \end{bmatrix}; \ M_s = \begin{bmatrix} M_{s11} & M_{s12} & M_{s13} \\ M_{s21} & M_{s22} & M_{s23} \\ M_{s31} & M_{s32} & M_{s33} \end{bmatrix};$$

$$K_s = \begin{bmatrix} K_{s11} & K_{s12} & K_{s13} \\ K_{s21} & K_{s22} & K_{s23} \\ K_{s31} & K_{s32} & K_{s33} \end{bmatrix} \tag{3.1.23a, b, c}$$

Because of cyclic symmetry,

$$\mathbf{u}_R = e^{j\psi_i} \mathbf{u}_L \quad \text{and} \quad \mathbf{f}_s = e^{j\psi_i} \mathbf{f}_{s-1}; \ j = \sqrt{-1} \tag{3.1.24a, b}$$

Using Equations (3.1.24 a, b), Equation (3.1.22) reduces to

$$M_t^c \ddot{\mathbf{u}}_1 + K_t^c \mathbf{u}_1 = \begin{bmatrix} \mathbf{0} \\ \mathbf{0} \end{bmatrix} \tag{3.1.25}$$

where

$$\mathbf{u}_1 = \begin{bmatrix} \mathbf{u}_L \\ \mathbf{u}_b \end{bmatrix} \tag{3.1.26}$$

$$K_t^c = \begin{bmatrix} K_{s11} + e^{j\psi_i} K_{s13} + K_{s33} + e^{-j\psi_i} K_{s31} & K_{s12} + e^{-j\psi_i} K_{s32} \\ K_{s21} + e^{j\psi_i} K_{s23} & K_{s22} \end{bmatrix} \tag{3.1.27}$$

and

$$M_t^c = \begin{bmatrix} M_{s11} + e^{j\psi_i} M_{s13} + M_{s33} + e^{-j\psi_i} M_{s31} & M_{s12} + e^{-j\psi_i} M_{s32} \\ M_{s21} + e^{j\psi_i} M_{s23} & M_{s22} \end{bmatrix} \tag{3.1.28}$$

It should be noted that the Equation (3.1.25) is written for sector number 1. Separating real and imaginary parts for the stiffness matrix,

$$K_t^c = K_{t,r\ell}^c + jK_{t,im}^c; \; j = \sqrt{-1} \tag{3.1.29}$$

where

$$K_{t,r\ell}^c = \begin{bmatrix} K_{s11} + K_{s33} + \cos \psi_i (K_{s13} + K_{s31}) & K_{s12} + K_{s32} \cos \psi_i \\ K_{s21} + K_{s23} \cos \psi_i & K_{s22} \end{bmatrix} \tag{3.1.30}$$

$$K_{t,im}^c = \begin{bmatrix} \sin \psi_i (K_{s13} - K_{s31}) & -K_{s32} \sin \psi_i \\ K_{s23} \sin \psi_i & 0 \end{bmatrix} \tag{3.1.31}$$

Similarly, separating real and imaginary parts for the mass matrix,

$$M_t^c = M_{t,r\ell}^c + jM_{t,im}^c; \;\; j = \sqrt{-1} \tag{3.1.32}$$

where

$$M_{t,r\ell}^c = \begin{bmatrix} M_{s11} + M_{s33} + \cos \psi_i (M_{s13} + M_{s31}) & M_{s12} + M_{s32} \cos \psi_i \\ M_{s21} + M_{s23} \cos \psi_i & M_{s22} \end{bmatrix} \tag{3.1.33}$$

$$M_{t,im}^c = \begin{bmatrix} \sin \psi_i (M_{s13} - M_{s31}) & -M_{s32} \sin \psi_i \\ M_{s23} \sin \psi_i & 0 \end{bmatrix} \tag{3.1.34}$$

Equations (3.1.29) and (3.1.32) provide detailed expressions for $M_t^c$ and $K_t^c$ in Equation (3.1.16). Similarly, detailed expressions for $\mathbf{u}_{r\ell}$ and $\mathbf{u}_{im}$ in Equation (3.1.19) are provided as follows.

$$\mathbf{u}_1 = \begin{bmatrix} \mathbf{u}_L \\ \mathbf{u}_b \end{bmatrix} = \begin{bmatrix} \mathbf{u}_{L,r\ell} \\ \mathbf{u}_b, r\ell \end{bmatrix} + j \begin{bmatrix} \mathbf{u}_{L,im} \\ \mathbf{u}_{b,im} \end{bmatrix} = \mathbf{u}_{r\ell} + j\mathbf{u}_{im} \tag{3.1.35}$$

Corresponding to Equation (3.1.20), double sectors are created as shown in Figure 3.1.2 where from Equation (3.1.24a),

$$\begin{bmatrix} \mathbf{u}_{R,r\ell} \\ \mathbf{u}_R, im \end{bmatrix} = \begin{bmatrix} \cos \psi_i & -\sin \psi_i \\ \sin \psi_i & \cos \psi_i \end{bmatrix} \begin{bmatrix} \mathbf{u}_{L,r\ell} \\ \mathbf{u}_L, im \end{bmatrix} \tag{3.1.36}$$

It should be noted that for unrepeated eigenvalues, interblade or intersector phase angles are 0 and 180 degrees. In this case, imaginary parts of mass and stiffness matrices, $M_{t,im}^c$ and $K_{t,im}^c$, are zero, and there is no coupling between these two sectors.

Eigenvalues/eigenvectors of the double sector in Equation (3.1.20) is written as

$$\begin{bmatrix} K_{t,r\ell}^c & -K_{t,im}^c \\ K_{t,im}^c & K_{t,r\ell}^c \end{bmatrix} \begin{bmatrix} \mathbf{v} \\ \mathbf{w} \end{bmatrix} = \omega^2 \begin{bmatrix} M_{t,r\ell}^c & -M_{t,im}^c \\ M_{t,im}^c & M_{t,r\ell}^c \end{bmatrix} \begin{bmatrix} \mathbf{v} \\ \mathbf{w} \end{bmatrix} \tag{3.1.37}$$

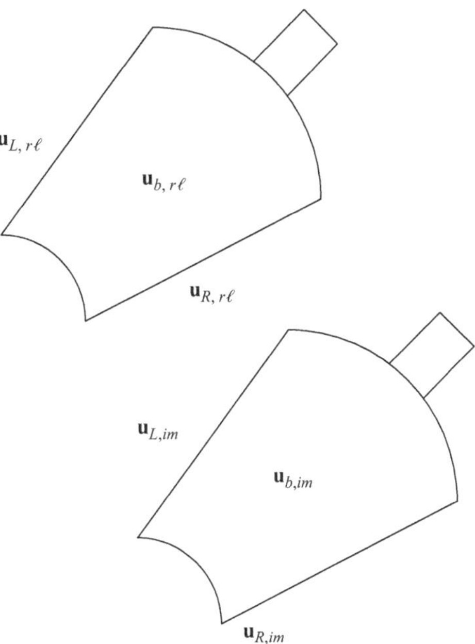


**Figure 3.1.2.** Double sector for analysis with real mass and stiffness matrices.

It can be easily seen that if $[\mathbf{v}^T \quad \mathbf{w}^T]^T$ is an eigenvector, $[-\mathbf{w}^T \quad \mathbf{v}^T]^T$ would also be an eigenvector for the repeated eigenvalue $\omega^2$. As a result, only one interblade phase angle for a repeated eigenvalue is considered (Ansys, 2012) because the double-sector formulation yields both the independent eigenvectors.

### 3.1.2 Brillouin Zone and Bloch's Theorem for Two-Dimensional Periodic Structures

A two-dimensional periodic structure is represented by an array of lattice points in space with repeated patterns (see Figure 3.1.3). A group of atoms is attached to each lattice point, and a crystal with periodic structure is obtained. In a two-dimensional structure, all lattice points can be defined by

$$\mathbf{r} = \mathbf{r}_j + \mathbf{r}_T \tag{3.1.38}$$

$$\mathbf{r}_T = \ell_1 \mathbf{a}_1 + \ell_2 \mathbf{a}_2 \tag{3.1.39}$$

where $\mathbf{a}_1$ and $\mathbf{a}_2$ are generating vectors, and $\ell_1$ and $\ell_2$ are integers. The parallelogram defined by generating vectors is described as primitive or unit or Weigner-Seitz primitive cell, which is the minimum volume cell that will fill the entire space after suitable translation operations. A general method to construct this primitive cell is described by Kittel (1996).

In a periodic lattice, physical properties are periodic with respect to the vector $\mathbf{r}_T$, Equation (3.1.39), that is,

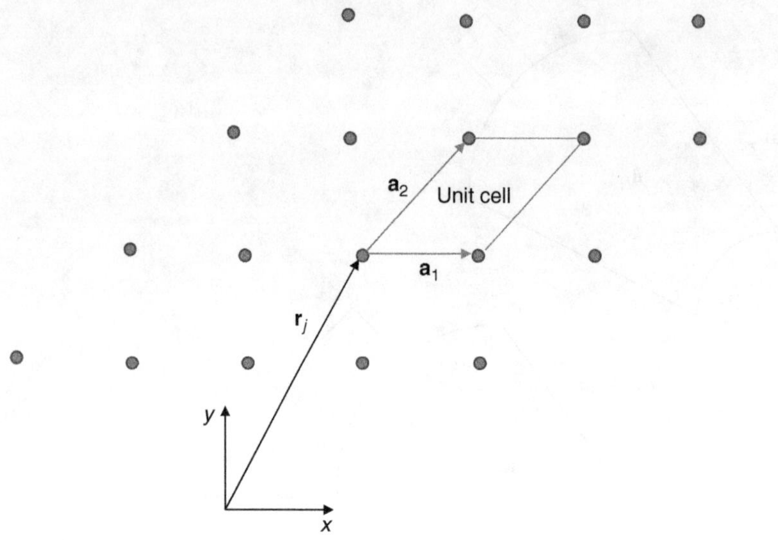


**Figure 3.1.3.**    Two-dimensional lattice.

$$f(\mathbf{r}+\mathbf{r}_T) = f(\mathbf{r}) \tag{3.1.40}$$

Fourier expansion can be written as (Cleland, 2003; Chen, 2005)

$$f(\mathbf{r}) = \sum_{\mathbf{k}} f_{\mathbf{k}}\, e^{j\mathbf{k}\cdot\mathbf{r}} \tag{3.1.41}$$

where $f_{\mathbf{k}}$ are Fourier coefficients, and

$$\mathbf{k} = p_1\mathbf{b}_1 + p_2\mathbf{b}_2; \ p_1 \quad \text{and} \quad p_2 \text{ integers} \tag{3.1.42}$$

$$\mathbf{a}_i \cdot \mathbf{b}_q = 2\pi\delta_{iq}; \ i = 1,2 \quad \text{and} \quad q = 1,2 \tag{3.1.43}$$

where $\delta_{iq}$ is the Kronecker delta function.

$$\delta_{iq} = \begin{cases} 1 & when \quad i = q \\ 0 & when \quad i \neq q \end{cases} \tag{3.1.44}$$

As an example, for a square lattice,

$$\mathbf{a}_1 = a\hat{\mathbf{x}};\ \mathbf{a}_2 = a\hat{\mathbf{y}} \tag{3.1.45}$$

$$\mathbf{b}_1 = \frac{2\pi}{a}\hat{\mathbf{x}};\ \mathbf{b}_2 = \frac{2\pi}{a}\hat{\mathbf{y}} \tag{3.1.46}$$

where $\hat{\mathbf{x}}$ and $\hat{\mathbf{y}}$ are unit vectors along x and y directions. Relationship (3.1.42) represents a periodic lattice, known as reciprocal lattice (see Figure 3.1.4). The parallelogram defined by generating vectors $\mathbf{b}_1$ and $\mathbf{b}_2$ is described as Weigner-Seitz unit

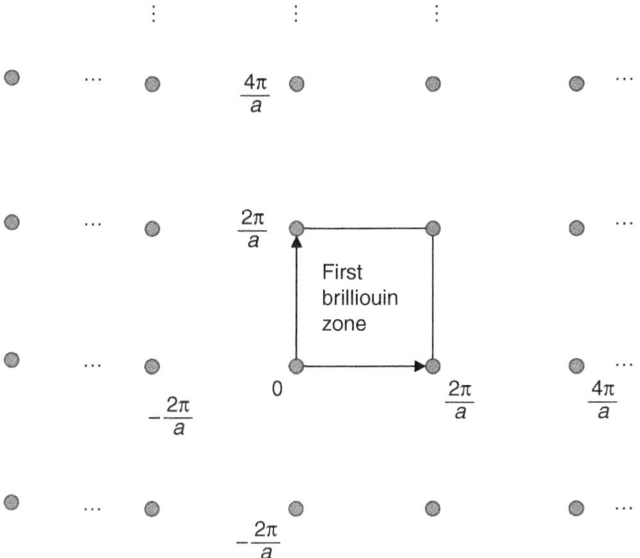


**Figure 3.1.4.**  Reciprocal square lattice.

cell, also known as Brillouin zone. The central cell in Figure 3.1.4 is known as the first Brillouin zone.

From Equations (3.1.39), (3.1.42), and (3.1.43),

$$\mathbf{k} \cdot \mathbf{r}_T = 2\pi(\ell_1 p_1 + \ell_2 p_2) \tag{3.1.47}$$

Therefore,

$$e^{j\mathbf{k} \cdot \mathbf{r}_T} = 1 \tag{3.1.48}$$

and the periodic property of the lattice is preserved, that is,

$$f(\mathbf{r} + \mathbf{r}_T) = \sum_{\mathbf{k}} f_{\mathbf{k}} e^{j\mathbf{k} \cdot (\mathbf{r} + \mathbf{r}_T)} = \sum_{\mathbf{k}} f_{\mathbf{k}} e^{j\mathbf{k} \cdot \mathbf{r}} e^{j\mathbf{k} \cdot \mathbf{r}_T} = f(\mathbf{r}) \tag{3.1.49}$$

According to Bloch's theorem, displacements corresponding to any eigenvector in any unit cell is related to those in unit cell number $j$ as

$$q(\mathbf{r}) = q(\mathbf{r}_j) e^{\mathbf{v}_{ip} \bullet (\mathbf{r} - \mathbf{r}_j)} = q(\mathbf{r}_j) e^{\mathbf{v}_{ip} \bullet (\ell \mathbf{a}_1 + p \mathbf{a}_2)} \tag{3.1.50}$$

where $\mathbf{v}_{ip}$ is the wave vector in the first Brillouin zone (Kittel, 1996), that is,

$$\mathbf{v}_{ip} = \mathbf{v}_{1i} + \mathbf{v}_{2p} \tag{3.1.51}$$

$$\mathbf{v}_{1i} = \frac{i}{n_{u1}} \mathbf{b}_1 \, ; \, i = 0,1,2,\ldots,n_{u1} - 1 \tag{3.1.52}$$

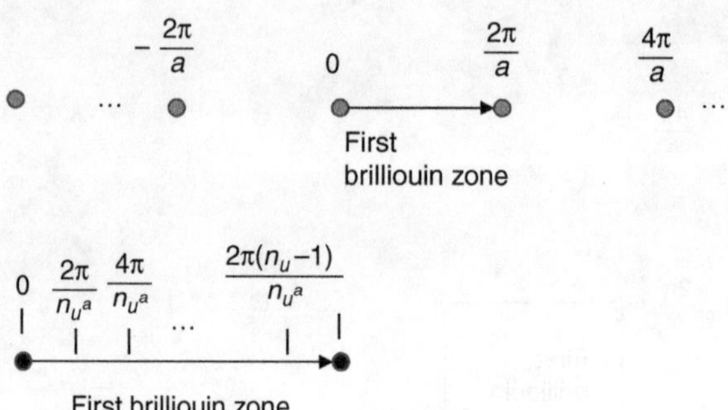


**Figure 3.1.5.**    Wave vectors and first Brillouin zone for one-dimensional periodic lattice.

$$\mathbf{v}_{2p} = \frac{p}{n_{u2}} \mathbf{b}_2; \quad p = 0, 1, 2, \ldots, n_{u2} - 1 \tag{3.1.53}$$

And, $n_{u1}$ and $n_{u2}$ are numbers of unit cells along directions of generating vectors $\mathbf{a}_1$ and $\mathbf{a}_2$. For a rotationally periodic structure like nanotubes, one of these numbers will be finite. For a periodic structure like graphene, $n_{u1} \to \infty$ and $n_{u2} \to \infty$. For a one-dimensional rotationally periodic structure, wave vectors in the first Brilliouin zone are shown in Figure 3.1.5.

For a unit cell of a two-dimensional undamped periodic structure (Figure 3.1.6), differential equations of motion are written as

$$M\ddot{\mathbf{q}} + K\mathbf{q} = \mathbf{f} \tag{3.1.54}$$

where $M$ and $K$ are mass and stiffness matrices, respectively. Components of displacement vector $\mathbf{q}$ and associated force vector $\mathbf{f}$ are shown in Figure 3.1.6, and described by Equations (3.1.55a) and (3.1.55b).

$$\mathbf{q} = \begin{bmatrix} \mathbf{q}_L \\ \mathbf{q}_R \\ \mathbf{q}_B \\ \mathbf{q}_T \\ \mathbf{q}_{LB} \\ \mathbf{q}_{LT} \\ \mathbf{q}_{RB} \\ \mathbf{q}_{RT} \\ \mathbf{q}_I \end{bmatrix} \quad \text{and} \quad \mathbf{f} = \begin{bmatrix} \mathbf{f}_L \\ \mathbf{f}_R \\ \mathbf{f}_B \\ \mathbf{f}_T \\ \mathbf{f}_{LB} \\ \mathbf{f}_{LT} \\ \mathbf{f}_{RB} \\ \mathbf{f}_{RT} \\ \mathbf{f}_I \end{bmatrix} \tag{3.1.55a, b}$$

Applying Bloch's theorem, Equation (3.1.50), the following relationship can be written:

$$\mathbf{q} = T\tilde{\mathbf{q}} \tag{3.1.56}$$

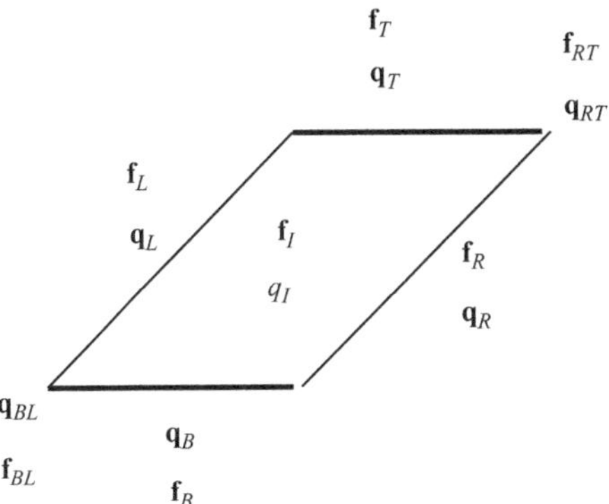


**Figure 3.1.6.** Unit cell of two-dimensional periodic structure.

where

$$
T = \begin{bmatrix}
I & 0 & 0 & 0 \\
e^{jk_{1i}} & 0 & 0 & 0 \\
0 & I & 0 & 0 \\
0 & e^{jk_{2p}} & 0 & 0 \\
0 & 0 & I & 0 \\
0 & 0 & e^{jk_{2p}} & 0 \\
0 & 0 & e^{jk_{1i}} & 0 \\
0 & 0 & e^{jk_{ip}} & 0 \\
0 & 0 & 0 & I
\end{bmatrix}
\quad \text{and} \quad
\tilde{\mathbf{q}} = \begin{bmatrix}
\mathbf{q}_L \\
\mathbf{q}_B \\
\mathbf{q}_{LB} \\
\mathbf{q}_I
\end{bmatrix}
\qquad (3.1.57\text{a, b})
$$

$$
k_{1i} = \mathbf{v}_{1i}.\mathbf{a}_1 = \frac{2\pi i}{n_{u1}} \qquad (3.1.58)
$$

$$
k_{2p} = \mathbf{v}_{2p}.\mathbf{a}_2 = \frac{2\pi p}{n_{u1}} \qquad (3.1.59)
$$

$$
k_{ip} = k_{1i} + k_{2p} \qquad (3.1.60)
$$

Substituting Equation (3.1.56) into Equation (3.1.54) and premultiplying both sides by $T^H$ (Phani, Woodhouse, and Fleck, 2006),

$$
\tilde{M}\ddot{\tilde{\mathbf{q}}} + \tilde{K}\tilde{\mathbf{q}} = \tilde{\mathbf{f}} \qquad (3.1.61)
$$

$$
\tilde{M} = T^H M T; \ \tilde{K} = T^H K T; \ \tilde{\mathbf{f}} = T^H \mathbf{f} \qquad (3.1.62\text{a, b, c})
$$

It should be noted that the reduced-order equation for a unit cell is similar to the cyclic symmetry analysis presented in section 3.1.1.

## 3.2     Reduced-Order Model: Frequency Mistuning

### 3.2.1     Subset of Nominal Modes/Modal Domain Analysis

This method (Yang and Griffin, 1999) is based on the fact that any mistuned mode can be represented as a linear combination of tuned modes. Further, the mass matrix is not changed, and the stiffness matrix is changed to simulate the changes in blade-alone frequencies due to mistuning, which is captured by appropriately varying the Young's moduli of elasticity of the blades.

Then the equations of motion for the full bladed disk can be written as:

$$M_t \ddot{\mathbf{x}} + K^{freq} \mathbf{x}(t) = 0 \tag{3.2.1}$$

where

$$K^{freq} = K_t + \delta K^{freq} \tag{3.2.2}$$

And, $M_t$ and $K_t$ are mass and stiffness matrices of the perfectly tuned system. The solution $\mathbf{x}$ is represented as a weighted sum of the modes of the nominal (tuned) bladed disk, that is,

$$\mathbf{x}(t) \approx \Phi_0 \mathbf{y}(t) \tag{3.2.3}$$

where the matrix $\Phi_0$ is composed of a set of tuned modes for the nominal bladed disk.

Substituting Equation (3.2.3) into Equation (3.2.1) and premultiplying with $\Phi_0^H$ (complex conjugate transpose of $\Phi_0$), the reduced-order equation of motion can be written as:

$$M_r^{SNM} \ddot{\mathbf{y}} + K_r^{SNM} \mathbf{y} = 0 \tag{3.2.4}$$

where

$$M_r^{SNM} = \Phi_0^H M_t \Phi_0 = I \tag{3.2.5}$$

$$K_r^{SNM} = \Phi_0^H K_t \Phi_0 + \Phi_0^H \delta K^{freq} \Phi_0 = \Lambda_t + \Phi_0^H \delta K^{freq} \Phi_0 \tag{3.2.6}$$

$$\delta K^{freq} = \begin{bmatrix} \delta K_1^{freq} & & & \\ & \delta K_2^{freq} & & \\ & & \ddots & \\ & & & \delta K_n^{freq} \end{bmatrix} \tag{3.2.7}$$

and $\Lambda_t$ is a diagonal matrix with square of tuned natural frequencies as its elements. Let $\varphi_{0,p,i}$ represent nodal displacements associated with the tuned mode, sector $p$, and the interblade phase angle $= 2\pi i / n$. Note that

$$\varphi_{0,p,i} = e^{j(p-1)\psi_i}\, \varphi_{0,1,i}; \; j = \sqrt{-1}; \tag{3.2.8}$$

where

$$\psi_i = \frac{2\pi i}{n} \qquad (3.2.9)$$

To compute the frequency mistuning term in Equation (3.2.6),

$$\Phi_0^H \, \delta K^{freq} \Phi_0 = \sum_{\ell=1}^{n} \Phi_0^H \, \delta K_\ell^b \Phi_0 \qquad (3.2.10)$$

where

$$\delta K_\ell^b = \begin{bmatrix} 0 & \cdots & 0 & \cdots & 0 \\ \cdots & \ddots & \cdots & \cdots & \cdots \\ 0 & \cdots & \delta K_\ell^{freq} & \cdots & 0 \\ \cdots & \cdots & \cdots & \ddots & \cdots \\ 0 & \cdots & 0 & \cdots & 0 \end{bmatrix} \qquad (3.2.11)$$

The $(\nu+1, \rho+1)$ element of $\Phi_0^H \, \delta K_\ell^b \Phi_0$ in Equation (3.2.10) is as follows:

$$
\begin{aligned}
\varphi_{0,1,\nu}^H \, \delta K_\ell^b \, e^{j\ell(\psi_\rho - \psi_\nu)} \varphi_{0,1,\rho} =\ & [\varphi_{0,1,\nu}^{R'} \quad \varphi_{0,1,\nu}^{I'}] \begin{bmatrix} \delta K_{\ell c} & -\delta K_{\ell s} \\ \delta K_{\ell s} & \delta K_{\ell c} \end{bmatrix} \begin{bmatrix} \varphi_{0,1,\rho}^{R} \\ \varphi_{0,1,\rho}^{I} \end{bmatrix} \\
& + j[-\varphi_{0,1,\nu}^{I'} \quad \varphi_{0,1,\nu}^{R'}] \begin{bmatrix} \delta K_{\ell c} & -\delta K_{\ell s} \\ \delta K_{\ell s} & \delta K_{\ell c} \end{bmatrix} \begin{bmatrix} \varphi_{0,1,\rho}^{R} \\ \varphi_{0,1,\rho}^{I} \end{bmatrix}
\end{aligned}
\qquad (3.2.12)
$$

where

$$\delta K_{\ell c} = \delta K_\ell^{freq} \cos(\ell(\psi_\rho - \psi_\nu)) \ \text{ and } \ \delta K_{\ell s} = \delta K_\ell^{freq} \sin(\ell(\psi_\rho - \psi_\nu)) \qquad (3.2.13a,\ b)$$

and $\varphi_{0,1,\rho}^{R}$ and $\varphi_{0,1,\rho}^{I}$ are real and imaginary parts of $\varphi_{0,1,\rho}$.

The eigenvalue problem associated with Equation (3.2.4) can be solved to get the mode shapes and natural frequencies of the mistuned bladed disk. The representation of actual geometric mistuning in terms of frequency mistuning (equivalent changes in Young's moduli of elasticity of blades) involves the following steps:

1. Determination of the natural frequency $(\omega_{bt})$ of the blade with average geometry and Young's modulus of elasticity $E_0$ clamped at base (Figure 3.2.1a).

   As a simple example, the average geometry is represented by the uniform thickness of a cantilever beam $b_t$. From the equivalent single degree of freedom model of a cantilever beam (Sinha, 2010),

$$\omega_{bt} = \sqrt{\frac{E_0 b_t^3 w}{4 \ell^3 m_{eq}}} \qquad (3.2.14)$$

where $m_{eq}$ is the equivalent mass of the cantilever beam.

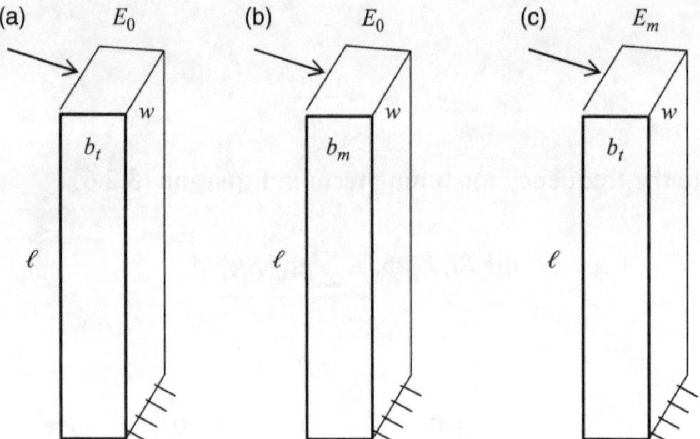


**Figure 3.2.1.**    a: Tuned blade, b: Geometrically mistuned blade, c: Blade with equivalent frequency mistuning.

2. Determination of the natural frequency $(\omega_{bm})$ of the mistuned blade with Young's modulus of elasticity $E_0$ clamped at base (Figure 3.2.1b).

   As a simple example, the geometric mistuning is represented by the variation in the thickness. With the mistuned thickness $= b_m$, mistuned frequencies are

$$\omega_{bm} = \sqrt{\frac{E_0 b_m^3 w}{4\ell^3 m_{eq}}} \tag{3.2.15}$$

3. Calculation of equivalent Young's modulus for a blade with average geometry such that the natural frequency of the blade is same as the natural frequency of the mistuned blade $(\omega_{bm})$, (Figure 3.2.1c). The equivalent Young's modulus can be calculated as:

$$E_m = E_0 \left( \frac{\omega_{bm}}{\omega_{bt}} \right)^2 \tag{3.2.16}$$

The finite element model of the bladed disk with the blades modeled as represented in Figure 3.2.1c is used to generate the mass and stiffness matrices of the mistuned bladed disk assembly.

## 3.2.2    Single-Family Mode Model

Feiner and Griffin (2002) developed a single-family reduced-order model, which they called fundamental model of mistuning (FMM). For a mistuned bladed rotor

$$[(K_t + \Delta K) - \omega_i^2 (M_t + \Delta M)]\varphi_i = 0 \tag{3.2.17}$$

where $\varphi_i$ and $\omega_i$ are ith mistuned mode and natural frequency, respectively. Let

$$\varphi_i = \sum_{m=0}^{n-1} \beta_{im}\varphi_m^t \tag{3.2.18}$$

where $\varphi_m^t$ is the mth tuned mode normalized with respect to the tuned mass matrix. Expressing mistuned mode as a linear combination of $n$ tuned modes,

$$\varphi_i = \Phi^t\beta_i \tag{3.2.19}$$

$$\Phi^t = \begin{bmatrix} \varphi_0^t & \varphi_1^t & \cdots & \varphi_{n-1}^t \end{bmatrix} \tag{3.2.20}$$

$$\beta_i = \begin{bmatrix} \beta_{i0} & \beta_{i1} & \cdots & \beta_{in-1} \end{bmatrix}^T \tag{3.2.21}$$

Substituting Equation (3.2.19) into Equation (3.2.17) and premultiplying by $\Phi^{tH}$,

$$[(\Lambda_t + \Delta\hat{K}) - \omega_i^2(I + \Delta\hat{M})]\beta_i = 0 \tag{3.2.22}$$

where

$$\Delta\hat{K} = \Phi^{tH}\Delta K\Phi^t \tag{3.2.23}$$

and

$$\Delta\hat{M} = \Phi^{tH}\Delta M\Phi^t \tag{3.2.24}$$

Equation (3.2.22) can be written as

$$(\Lambda_t + \hat{A}_i)\beta_i = \omega_i^2\beta_i \tag{3.2.25}$$

where

$$\hat{A}_i = \Delta\hat{K} - \omega_i^2\Delta\hat{M} \tag{3.2.26}$$

Assume that $\Phi^t$ is composed of a single family of blade-dominated modes. In this case, all frequencies $\omega_i$ will be closely spaced around an average frequency $\omega_{av}$, and it will be assumed that

$$\omega_i \approx \omega_{av} \tag{3.2.27}$$

$$\hat{A}_i \approx \hat{A} = \Delta\hat{K} - \omega_{av}^2\Delta\hat{M} \tag{3.2.28}$$

Equation (3.2.25) is approximated as

$$(\Lambda_t + \hat{A})\beta_i = \omega_i^2\beta_i \tag{3.2.29}$$

Now,

$$\hat{A} = \sum_{s=0}^{n-1} \hat{A}_s \tag{3.2.30}$$

where $\hat{A}_s$ is the contribution from sector number $s$. From Equations (3.2.28) and (3.2.30),

$$\hat{A}_s = \Delta\hat{K}_s - \omega_{av}^2\,\Delta\hat{M}_s \tag{3.2.31}$$

Because of block-diagonal structures of $\Delta K$ and $\Delta M$, Equations (3.2.23) and (3.2.24) yield

$$\Delta\hat{K}_s = \Phi_s^{t^H}\,\Delta K_s\,\Phi_s^t \tag{3.2.32}$$

$$\Delta\hat{M}_s = \Phi_s^{t^H}\,\Delta M_s\,\Phi_s^t \tag{3.2.33}$$

where $\Delta K_s$ and $\Delta M_s$ are deviations in stiffness and mass matrices for sector number $s$. Substituting Equations (3.2.32) and (3.2.33) into Equation (3.2.31),

$$\hat{A}_s = \Phi_s^{t^H}(\Delta K_s - \omega_{av}^2\,\Delta M_s)\Phi_s^t \tag{3.2.34}$$

A single element of $\hat{A}_s$, for example, the element in column number $i$ and row number $m$, can be written as

$$\hat{A}_{s,mi} = \varphi_{s,m}^{t^H}(\Delta K_s - \omega_{av}^2\,\Delta M_s)\varphi_{s,i}^t \tag{3.2.35}$$

$$\varphi_{s,m}^t = e^{jsm\psi}\,\varphi_{0,m}^t \tag{3.2.36}$$

where

$$\psi = \frac{2\pi}{n} \tag{3.2.37}$$

and $\varphi_{0,m}^t$ is the tuned mode vector for 0th sector. Substituting Equation (3.2.36) into Equation (3.2.35),

$$\hat{A}_{s,mi} = e^{js(i-m)\psi}\,\varphi_{0,m}^{t^H}(\Delta K_s - \omega_{av}^2\,\Delta M_s)\varphi_{0,i}^t \tag{3.2.38}$$

Because it is assumed that most of the strain energies are in blades, sector modes are all identical. In this case, Equation (3.2.38) can be approximated as

$$\hat{A}_{s,mi} = \frac{\omega_{av}^2}{n\omega_{t,b}^2}\,e^{js(i-m)\psi}\,\mathbf{b}_t^H(\Delta K_{s,b} - \omega_{av}^2\,\Delta M_{s,b})\mathbf{b}_t \tag{3.2.39}$$

where $\mathbf{b}_t$ is the modal vector of the clamped tuned blade alone normalized with respect to the blade mass matrix, $\omega_{t,b}$ is the natural frequency of tuned blade, and $\Delta K_{s,b}$ and $\Delta M_{s,b}$ are deviations in mass and stiffness matrices for blade number s alone. The factor $\omega_{av}^2/(n\omega_{t,b}^2)$ has been introduced to scale the blade mode terms so that they have approximately the same strain energy as that in the sector.

For a clamped mistuned blade number $s$, natural frequency/modal vector equation is

$$[(K_{t,b} + \Delta K_{s,b}) - \omega_{s,b}^2 (M_{t,b} + \Delta M_{s,b})]\mathbf{b}_s = 0 \qquad (3.2.40)$$

where $\mathbf{b}_s$ is the modal vector of the clamped blade number s, and $K_{t,b}$ and $M_{t,b}$ are mass and stiffness matrices for the tuned blade. It is assumed that the blade's modal vector does not change significantly due to mistuning, that is,

$$\mathbf{b}_s \approx \mathbf{b}_t \qquad (3.2.41)$$

Premultiplying Equation (3.2.40) by $\mathbf{b}_t^H$ and using the approximation (3.2.41),

$$\mathbf{b}_t^H [(K_{t,b} + \Delta K_{s,b}) - \omega_{s,b}^2 (M_{t,b} + \Delta M_{s,b})]\mathbf{b}_t = 0 \qquad (3.2.42)$$

Note that

$$\mathbf{b}_t^H K_{t,b}\, \mathbf{b}_t = \omega_{t,b}^2 \qquad (3.2.43)$$

$$\mathbf{b}_t^H M_{t,b}\, \mathbf{b}_t = 1 \qquad (3.2.44)$$

From Equations (3.2.42) – (3.2.44),

$$\mathbf{b}_t^H (\Delta K_{s,b} - \omega_{s,b}^2 \Delta M_{s,b})\mathbf{b}_t = \omega_{s,b}^2 - \omega_{t,b}^2 \qquad (3.2.45)$$

Further,

$$\omega_{s,b}^2 - \omega_{t,b}^2 = (\omega_{s,b} + \omega_{t,b})(\omega_{s,b} - \omega_{t,b}) \approx 2\omega_{t,b}^2 \Delta_f \omega_{s,b} \qquad (3.2.46)$$

where

$$\Delta_f \omega_{s,b} = \frac{(\omega_{s,b} - \omega_{t,b})}{\omega_{t,b}} \qquad (3.2.47)$$

It is assumed that

$$\mathbf{b}_t^H (\Delta K_{s,b} - \omega_{s,b}^2 \Delta M_{s,b})\mathbf{b}_t \approx \mathbf{b}_t^H (\Delta K_{s,b} - \omega_{av}^2 \Delta M_{s,b})\mathbf{b}_t \qquad (3.2.48)$$

From Equations (3.2.45)–(3.2.47),

$$\mathbf{b}_t^H (\Delta K_{s,b} - \omega_{av}^2 \Delta M_{s,b})\mathbf{b}_t = 2\omega_{t,b}^2 \Delta_f \omega_{s,b} \qquad (3.2.49)$$

Substituting Equation (3.2.49) into Equation (3.2.39),

$$\hat{A}_{s,mi} = \frac{2\omega_{av}^2}{n} e^{js(i-m)\psi} \Delta_f \omega_{s,b} \qquad (3.2.50)$$

From Equations (3.2.30) and (3.2.50),

$$\hat{A}_{mi} = \sum_{s=0}^{n-1} \hat{A}_{s,mi} = \frac{2\omega_{av}^2}{n} \sum_{s=0}^{n-1} e^{jsp\psi} \Delta_f \omega_{s,b} \tag{3.2.51}$$

where $p = i - m$. The range of $p$ will always be from 0 to $n$-1 because

$$p = \begin{cases} i - m & \text{for } i \geq m \\ i - m + n & \text{for } i < m \end{cases} \tag{3.2.52}$$

From Equation (3.2.51),

$$\hat{A}_{mi} = 2\omega_{av}^2 \bar{\omega}_p \tag{3.2.53}$$

where $\bar{\omega}_p$ is the pth discrete Fourier transform (Oppenheim and Shaeffer, 1975) of the blade frequency deviation $\Delta_f \omega_{s,b}$, defined as

$$\bar{\omega}_p = \frac{1}{N} \sum_{s=0}^{n-1} e^{jsp\psi} \Delta_f \omega_{s,b} \tag{3.2.54}$$

From Equation (3.2.53),

$$\hat{A} = 2\omega_{av}^2 \bar{\Omega} \tag{3.2.55}$$

where

$$\bar{\Omega} = \begin{bmatrix} \bar{\omega}_0 & \bar{\omega}_1 & . & , & \bar{\omega}_{n-1} \\ \bar{\omega}_{n-1} & \bar{\omega}_0 & . & . & \bar{\omega}_{n-2} \\ . & & . & . & . \\ . & & . & . & . \\ \bar{\omega}_1 & \bar{\omega}_2 & . & . & \bar{\omega}_0 \end{bmatrix} \tag{3.2.56}$$

Using Equation (3.2.55), Equation (3.2.29) can be expressed as

$$(\Lambda_t + 2\omega_{av}^2 \bar{\Omega})\beta_i = \omega_i^2 \beta_i \tag{3.2.57}$$

Equation (3.2.57) is the eigenvalue/vector problem for the SFFM. From the knowledge of tuned mode frequencies $\Lambda_t$ and frequency mistuning $\Delta_f \omega_{s,b}$, mistuned frequencies $\omega_i$ and mistuned mode shapes $\beta_i$ can be computed.

The structure of this SFMM model is same as that of the modal equations of basic spring mass model shown in Figure 1.1.1. Equation (3.2.50) is now rewritten for deviations in stiffness, $\delta k_s$, Equation (1.1.2). From Equation (2.2.5) with zero forcing function, zero damping, and $\Delta M = 0$,

$$(\Omega^2 + \Phi^H \Delta K \Phi)\alpha_t = \omega^2 \alpha_t \tag{3.2.58}$$

Then, the element in column number $i$ and row number $m$ of the matrix $\Phi^H \Delta K \Phi$ can be written as

$$\frac{1}{nm_t} \sum_{s=1}^{n} e^{j(s-1)p\psi} \delta k_s = \frac{1}{m_t} \delta \bar{k}_p \tag{3.2.59}$$

where p is defined by Equation (3.2.52) and $\delta \bar{k}_p$ is the pth discrete Fourier transform (DFT) of the blade stiffness deviation $\delta k_s$, defined as

$$\delta \bar{k}_p = \frac{1}{n} \sum_{s=1}^{n} e^{j(s-1)p\psi} \, \delta k_s \tag{3.2.60}$$

Therefore, Equation (3.2.58) can be written as

$$(\Omega^2 + \frac{1}{m_t} \bar{\Omega}_k)\alpha_t = \omega^2 \alpha_t \tag{3.2.61}$$

$$\bar{\Omega}_k = \begin{bmatrix} \delta \bar{k}_0 & \delta \bar{k}_1 & . & , & \delta \bar{k}_{n-1} \\ \delta \bar{k}_{n-1} & \delta \bar{k}_0 & . & . & \delta \bar{k}_{n-2} \\ . & & . & . & . \\ . & & . & . & . \\ \delta \bar{k}_1 & \delta \bar{k}_2 & . & . & \delta \bar{k}_0 \end{bmatrix} \tag{3.2.62}$$

Hence, Equations (3.2.57) and (3.2.61) are similar.

### 3.2.3 Identification of Frequency Mistuning from Measured Data

From Equation (3.2.22),

$$[(I + \Delta \hat{M})^{-1}(\Lambda_t + \Delta \hat{K}) - \omega_i^2 I]\beta_i = 0 \tag{3.2.63}$$

Feiner and Griffin (2004a) made following approximations by neglecting second- and higher-order terms:

$$(I + \Delta \hat{M})^{-1}(\Lambda_t + \Delta \hat{K}) \approx (I - \Delta \hat{M})(\Lambda_t + \Delta \hat{K}) \approx \Lambda_t + \hat{B} \tag{3.2.64}$$

where

$$\hat{B} = \Delta \hat{K} - \Delta \hat{M} \Lambda_t \tag{3.2.65}$$

Substituting Equation (3.2.64) into Equation (3.2.63),

$$(\Lambda_t + \hat{B})\beta_i = \omega_i^2 \beta_i \tag{3.2.66}$$

Representing $\hat{B}$ as sum of contributions from each sector,

$$\hat{B} = \sum_{s=1}^{n} \hat{B}_s \tag{3.2.67}$$

A single element of $\hat{B}_s$, for example, the element in column number $i$ and row number $m$, can be written as

$$\hat{B}_{s,mi} = \boldsymbol{\varphi}_{s,m}^{t\,H}(\Delta K_s - \omega_i^2 \Delta M_s)\boldsymbol{\varphi}_{s,i}^{t} \tag{3.2.68}$$

where $\Delta K_s$ and $\Delta M_s$ are deviations in stiffness and mass matrices of sector number $s$. Because of Equation (3.2.36),

$$\hat{B}_{s,mi} = e^{j(i-m)\psi}\boldsymbol{\varphi}_{0,m}^{t\,H}(\Delta K_s - \omega_i^2 \Delta M_s)\boldsymbol{\varphi}_{0,i}^{t} \tag{3.2.69}$$

Feiner and Griffin (2004a) developed this analysis for an isolated family of tuned modes, and replaced various sector modes by an average sector mode $\overline{\varphi}_0^t$. They approximated Equation (3.2.69) in an ad hoc manner by

$$\hat{B}_{s,mi} = \frac{\omega_m^t \omega_i^t}{\omega_{av}^2} e^{j(i-m)\psi}\overline{\boldsymbol{\varphi}}_{0,m}^{t\,H}(\Delta K_s - \omega_i^2 \Delta M_s)\boldsymbol{\varphi}_{0,n}^{t} \tag{3.2.70}$$

where $\omega_{av}$ is the average tuned mode frequency. Then, they considered tuned disks with sectors being one of the mistuned sectors, and again made many ad hoc approximations to get

$$\hat{B}_{s,mi} = \frac{2\omega_m^t \omega_i^t}{n} e^{j(i-m)\psi}\Delta\omega_{s,bd} \tag{3.2.71}$$

where

$$\Delta\omega_{s,bd} = \frac{\omega_{s,bd}^t - \omega_{av}}{\omega_{av}} \tag{3.2.72}$$

where $\omega_s^t$ is the natural frequency of the sector number $s$ corresponding to average sector mode. Adding contributions from each sector,

$$\hat{B}_{mi} = \sum_{s=0}^{n-1} \hat{B}_{s,mi} = 2\omega_m^t \,\overline{\omega}_{i-m,bd}\, \omega_i^t \tag{3.2.73}$$

where $\overline{\omega}_{i-m,bd}$ is the discrete Fourier transform of $\Delta\omega_{s,bd}$, that is,

$$\overline{\omega}_{i-m,bd} = \frac{1}{n}\sum_{s=0}^{n-1} e^{js(i-m)2\pi/n}\Delta\omega_{s,bd} \tag{3.2.74}$$

Equation (3.2.73) leads to

$$\hat{B} = 2\Omega_t \overline{\Omega}_{bd}\Omega_t \tag{3.2.75}$$

where

$$\Omega_t = \Lambda_t^{1/2} = diag[\omega_0' \quad \omega_1' \quad \cdots \quad \omega_{n-1}'] \tag{3.2.76}$$

and

$$\bar{\Omega}_{bd} = \begin{bmatrix} \bar{\omega}_{0,bd} & \bar{\omega}_{1,bd} & \cdots & \bar{\omega}_{n-1,bd} \\ \bar{\omega}_{n-1,bd} & \bar{\omega}_{0,bd} & \cdots & \bar{\omega}_{n-2,bd} \\ \vdots & \vdots & & \vdots \\ \bar{\omega}_{1,bd} & \bar{\omega}_{2,bd} & \cdots & \bar{\omega}_{0,bd} \end{bmatrix} \tag{3.2.77}$$

Substituting Equations (3.2.75) and (3.2.76) into Equation (3.2.66),

$$(\Omega_t^2 + 2\Omega_t\bar{\Omega}_{bd}\Omega_t)\beta_i = \omega_i^2\beta_i \tag{3.2.78}$$

The goal is to estimate $\bar{\Omega}_{bd}$ from the measurements of mistuned modes $\beta_i$ and mistuned frequencies $\omega_i$. Equation (3.2.78) can be rearranged as

$$2\Omega_t\bar{\Omega}_{bd}\mathbf{g}_i = (-\Omega_t^2 + \omega_i^2 I)\beta_i \tag{3.2.79}$$

where

$$\gamma_i = \Omega_t\beta_i \tag{3.2.80}$$

Next, it has been shown that

$$\bar{\Omega}\gamma_i = \Gamma_i\bar{\omega}_{bd} \tag{3.2.81}$$

where

$$\bar{\omega}_{bd} = [\bar{\omega}_{0,bd} \quad \bar{\omega}_{1,bd} \quad \cdots \quad \bar{\omega}_{n-1,bd}] \tag{3.2.82}$$

and

$$\Gamma_i = \begin{bmatrix} \gamma_{i0} & \gamma_{i1} & \cdots & \gamma_{i(n-1)} \\ \gamma_{i1} & \gamma_{i2} & \cdots & \gamma_{i0} \\ \vdots & \vdots & & \vdots \\ \gamma_{i(n-1)} & \gamma_{i0} & \cdots & \gamma_{i(n-2)} \end{bmatrix} \tag{3.2.83}$$

Therefore, from Equations (3.2.79) and (3.2.81),

$$2\Omega_t\Gamma_j\bar{\omega}_{bd} = \mathbf{r}_i \tag{3.2.84}$$

where

$$\mathbf{r}_i = (-\Omega_t^2 + \omega_i^2 I)\beta_i \tag{3.2.85}$$

With many measured mistuned modes, Equation (3.2.84) leads to

$$L\bar{\omega} = \mathbf{r}_{ai} \qquad (3.2.86)$$

where

$$L = \begin{bmatrix} 2\Omega_i\Gamma_0 \\ 2\Omega_i\Gamma_1 \\ \vdots \\ 2\Omega_i\Gamma_m \end{bmatrix} \quad \text{and} \quad \mathbf{r}_{ai} = \begin{bmatrix} \mathbf{r}_0 \\ \mathbf{r}_1 \\ \vdots \\ \mathbf{r}_p \end{bmatrix} \qquad (3.2.87)$$

Assuming that the matrix is of full rank, the least square error (Strang, 1988) solution is

$$\bar{\omega}_{bd} = (L^T L)^{-1} L^T \mathbf{r}_{ai} \qquad (3.2.88)$$

It is known that the least square error reduces the effects of measurement errors. Finally, frequency deviation of each sector is obtained by the inverse discrete Fourier transform:

$$\Delta\omega_{s,bd} = \sum_{p=0}^{n-1} e^{jsp2\pi/n}\,\bar{\omega}_{p,bd} \qquad (3.2.89)$$

Feiner and Griffin (2004a, 2004b) have provided many examples of application of this identification process. They have also provided formulation in which natural frequencies for the relevant family of tuned modes are also estimated.

## 3.3    Reduced-Order Model: Component Mode Synthesis

In the CMS approach (Craig, 1981), blades and disk are modeled separately, and then assembled together. This technique has similarity with the receptance approach (Rao, 1991), which was one of the first methods used to study vibration characteristics of a bladed disk assembly. Bladh, Castanier, and Pierre (2001) have used the CMS approach to develop the reduced-order model for a bladed disk with frequency mistuning. Brown (2008) and Beck (2010) have used the CMS approach to develop reduced-order models for a bladed disk with geometric mistuning. Here, fundamentals of development of the CMS-based reduced-order models will be presented. Numerical results can be found in the references: Bladh et al. (2001), Brown (2008), and Beck (2010).

Consider a bladed disk with $n$ blades, for example, Figure 3.3.1 with $n = 12$.

### Disk

Let the displacement vector of the interior degrees of freedom of the disk be $\mathbf{u}_i$, and that for boundary nodes common with blade number j be $\mathbf{u}_{b_j}$ (Figure 3.3.1), where $j = 1, 2, 3, \ldots, n$. Let the mass and stiffness matrix of the disk be

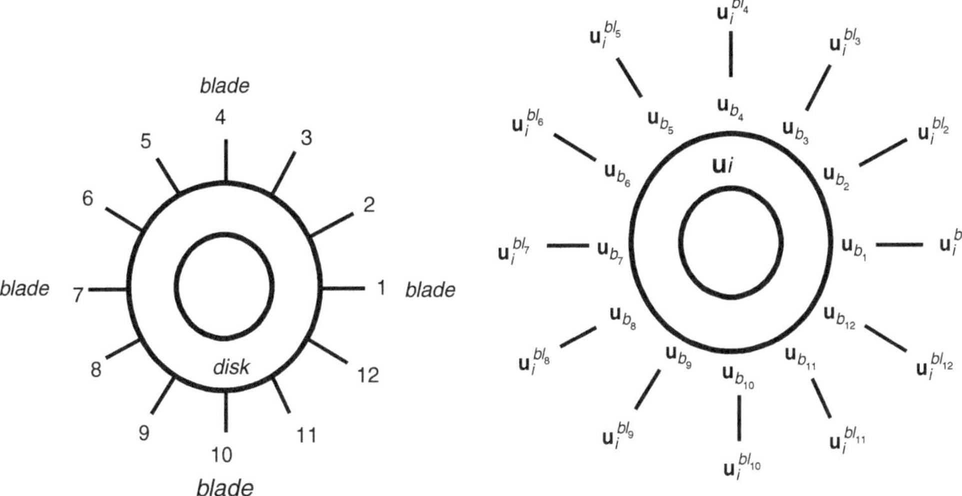

**Figure 3.3.1.** A bladed disk with twelve blades.

$$M_d = \begin{bmatrix} M_{b_1 b_1} & M_{b_1 b_2} & \cdots & M_{b_1 b_n} & M_{b_1 i} \\ M_{b_2 b_1} & M_{b_2 b_2} & \cdots & M_{b_2 b_n} & M_{b_2 i} \\ \vdots & \vdots & \ddots & \vdots & \vdots \\ M_{b_n b_1} & M_{b_n b_2} & \cdots & M_{b_n b_n} & M_{b_n i} \\ M_{i b_1} & M_{i b_2} & \cdots & M_{i b_n} & M_{ii} \end{bmatrix} \tag{3.3.1}$$

$$K_d = \begin{bmatrix} K_{b_1 b_1} & K_{b_1 b_2} & \cdots & K_{b_1 b_n} & K_{b_1 i} \\ K_{b_2 b_1} & K_{b_2 b_2} & \cdots & K_{b_2 b_n} & K_{b_2 i} \\ \vdots & \vdots & \ddots & \vdots & \vdots \\ K_{b_n b_1} & K_{b_n b_2} & \cdots & K_{b_n b_n} & K_{b_n i} \\ K_{i b_1} & K_{i b_2} & \cdots & K_{i b_n} & K_{ii} \end{bmatrix} \tag{3.3.2}$$

Free vibration of the disk will be described by

$$M_d \ddot{\mathbf{u}}_d + K_d \mathbf{u}_d = \mathbf{0} \tag{3.3.3}$$

where

$$\mathbf{u}_d^T = \begin{bmatrix} \mathbf{u}_{b_1}^T & \mathbf{u}_{b_2}^T & \cdots & \mathbf{u}_{b_n}^T & \mathbf{u}_i^T \end{bmatrix} \tag{3.3.4}$$

## Constrained Static Modes for Disk

Let the force vector of the interior degrees of freedom of the disk be $\mathbf{f}_i$, and that for boundary nodes common with blade number j be $\mathbf{f}_{b_j}$, where $j = 1, 2, 3, \ldots, n$. Then, static equilibrium equations will be given by

$$\begin{bmatrix} K_{b_1b_1} & K_{b_1b_2} & \cdots & K_{b_1b_n} & K_{b_1i} \\ K_{b_2b_1} & K_{b_2b_2} & \cdots & K_{b_2b_n} & K_{b_2i} \\ \vdots & \vdots & \ddots & \vdots & \vdots \\ K_{b_nb_1} & K_{b_nb_2} & \cdots & K_{b_nb_n} & K_{b_ni} \\ K_{ib_1} & K_{ib_2} & \cdots & K_{ib_n} & K_{ii} \end{bmatrix} \begin{bmatrix} \mathbf{u}_{b_1} \\ \mathbf{u}_{b_2} \\ \vdots \\ \mathbf{u}_{b_n} \\ \mathbf{u}_i \end{bmatrix} = \begin{bmatrix} \mathbf{f}_{b_1} \\ \mathbf{f}_{b_2} \\ \vdots \\ \mathbf{f}_{b_n} \\ \mathbf{f}_i \end{bmatrix} \tag{3.3.5}$$

With $\mathbf{f}_i = 0$,

$$K_{ib_1}\mathbf{u}_{b_1} + K_{ib_2}\mathbf{u}_{b_2} + \cdots + K_{ib_n}\mathbf{u}_{b_n} + K_{ii}\mathbf{u}_i = \mathbf{0} \tag{3.3.6}$$

or,

$$\mathbf{u}_i = \Phi_{cb_1}\mathbf{u}_{b_1} + \Phi_{cb_2}\mathbf{u}_{b_2} + \cdots + \Phi_{cb_n}\mathbf{u}_{b_n} \tag{3.3.7}$$

where

$$\Phi_{cb_j} = -K_{ii}^{-1}K_{ib_j}; \quad j = 1, 2, \ldots, n \tag{3.3.8}$$

Let there be unit displacement of $\mathbf{u}_{b_j}$ at a time. Further, all other boundary displacements are constrained to be zeros. In this case, $n$ sets of constrained displacement matrix can be obtained using Equation (3.3.6) as follows:

$$\begin{bmatrix} \mathbf{u}_{b_1} \\ \mathbf{u}_{b_2} \\ \vdots \\ \mathbf{u}_{b_n} \\ \mathbf{u}_i \end{bmatrix} = \begin{bmatrix} I_{b_1} \\ 0 \\ \vdots \\ 0 \\ \Phi_{cb_1} \end{bmatrix}, \begin{bmatrix} 0 \\ I_{b_2} \\ \vdots \\ 0 \\ \Phi_{cb_2} \end{bmatrix}, \ldots, \begin{bmatrix} 0 \\ 0 \\ \vdots \\ I_{b_n} \\ \Phi_{cb_n} \end{bmatrix} \tag{3.3.9}$$

## Constrained Vibratory Modes of Disk

Let displacements at blade nodes be zero, that is,

$$\mathbf{u}_{b_1} = 0, \ \mathbf{u}_{b_2} = 0, \ldots, \mathbf{u}_{b_n} = 0 \tag{3.3.10}$$

In this case, Equation (3.3.3) leads to following differential equations of motion

$$M_{ii}\ddot{\mathbf{u}}_i + K_{ii}\mathbf{u}_i = \mathbf{0} \tag{3.3.11}$$

The corresponding modal matrix is written as

$$K_{ii}\Phi_v = M_{ii}\Phi_v\Lambda_d \tag{3.3.12}$$

where

$$\Phi_v^T K_{ii} \Phi_v = \Lambda_d \quad \text{and} \quad \Phi_v^T M_{ii} \Phi_v = I_{d_i} \tag{3.3.13}$$

Here, $\Lambda_d$ is a diagonal matrix containing eigenvalues, and the matrix $\Phi_v$ is composed of eigenvectors.

## Model Reduction for Disk

Considering only $nv_d$ vibratory modes, the following approximation is made:

$$\mathbf{u}_d = \Psi_d \alpha_d \tag{3.3.14}$$

where

$$\alpha_d^T = \begin{bmatrix} \alpha_{b_1}^T & \alpha_{b_2}^T & \cdots & \alpha_{b_n}^T & \alpha_i^T \end{bmatrix} \tag{3.3.15}$$

and

$$\Psi_d = \begin{bmatrix} I_{b_1} & 0 & \cdots & 0 & 0 \\ 0 & I_{b_2} & \cdots & 0 & 0 \\ \vdots & \vdots & \ddots & \vdots & \vdots \\ 0 & 0 & \cdots & I_{b_n} & 0 \\ \Phi_{cb_1} & \Phi_{cb_2} & \cdots & \Phi_{cb_n} & \Phi_v \end{bmatrix} \tag{3.3.16}$$

Note that the dimension of $\Phi_v$ is $nv_d$. Let the dimension of $\alpha_{b_i}$ be $n_{b_i}$. Then, the dimension of $\alpha_d$, Equation (3.3.15), will be $nv_d + \sum_{i=1}^{n} n_{b_i}$.

Substituting Equation (3.3.14) into Equation (3.3.3),

$$\hat{M}_d \ddot{\alpha}_d + \hat{K}_d \alpha_d = 0 \tag{3.3.17}$$

where

$$\hat{M}_d = \Psi_d^T M_d \Psi_d \tag{3.3.18}$$

$$\hat{K}_d = \Psi_d^T K_d \Psi_d \tag{3.3.19}$$

$$\hat{K}_d = \begin{bmatrix} K_{b_1 b_1} + \Phi_{cb_1}^T K_{ib_1} & K_{b_1 b_2} + \Phi_{cb_1}^T K_{ib_2} & \cdots & K_{b_1 b_n} + \Phi_{cb_1}^T K_{ib_n} & 0 \\ K_{b_2 b_1} + \Phi_{cb_2}^T K_{ib_1} & K_{b_2 b_2} + \Phi_{cb_2}^T K_{ib_2} & \cdots & K_{b_2 b_n} + \Phi_{cb_2}^T K_{ib_n} & 0 \\ \vdots & \vdots & \ddots & \vdots & \vdots \\ K_{b_n b_1} + \Phi_{cb_n}^T K_{ib_1} & K_{b_n b_2} + \Phi_{cb_n}^T K_{ib_2} & \cdots & K_{b_n b_n} + \Phi_{cb_n}^T K_{ib_n} & 0 \\ 0 & 0 & \cdots & 0 & \Phi_v^T K_{ii} \Phi_v \end{bmatrix} \tag{3.3.20}$$

$$\hat{M}_d = \begin{bmatrix} M_{b_1 b_1} + \Phi_{cb_1}^T M_{ib_1} + M_{b_1 i}\Phi_{cb_1} + \Phi_{cb_1}^T M_{ii}\Phi_{cb_1} & M_{b_1 b_2} + \Phi_{cb_1}^T M_{ib_2} + M_{b_1 i}\Phi_{cb_2} + \Phi_{cb_1}^T M_{ii}\Phi_{cb_2} & \cdots & M_{b_1 b_n} + \Phi_{cb_1}^T M_{ib_n} + M_{b_1 i}\Phi_{cb_n} + \Phi_{cb_1}^T M_{ii}\Phi_{cb_n} & M_{b_1 i}\Phi_v + \Phi_{cb_1}^T M_{ii}\Phi_v \\[2ex] M_{b_2 b_1} + \Phi_{cb_2}^T M_{ib_1} + M_{b_2 i}\Phi_{cb_1} + \Phi_{cb_2}^T M_{ii}\Phi_{cb_1} & M_{b_2 b_2} + \Phi_{cb_2}^T M_{ib_2} + M_{b_2 i}\Phi_{cb_2} + \Phi_{cb_2}^T M_{ii}\Phi_{cb_2} & \cdots & M_{b_2 b_n} + \Phi_{cb_2}^T M_{ib_n} + M_{b_2 i}\Phi_{cb_n} + \Phi_{cb_2}^T M_{ii}\Phi_{cb_n} & M_{b_2 i}\Phi_v + \Phi_{cb_2}^T M_{ii}\Phi_v \\[2ex] \vdots & \vdots & \ddots & \vdots & \vdots \\[2ex] M_{b_n b_1} + \Phi_{cb_n}^T M_{ib_1} + M_{b_n i}\Phi_{cb_1} + \Phi_{cb_n}^T M_{ii}\Phi_{cb_1} & M_{b_n b_2} + \Phi_{cb_n}^T M_{ib_2} + M_{b_n i}\Phi_{cb_2} + \Phi_{cb_n}^T M_{ii}\Phi_{cb_2} & \cdots & M_{b_n b_n} + \Phi_{cb_n}^T M_{ib_n} + M_{b_n i}\Phi_{cb_n} + \Phi_{cb_n}^T M_{ii}\Phi_{cb_n} & M_{b_n i}\Phi_v + \Phi_{cb_n}^T M_{ii}\Phi_v \\[2ex] \Phi_v^T M_{ii}\Phi_{cb_1} + \Phi_v^T M_{ib_1} & \Phi_v^T M_{ii}\Phi_{cb_2} + \Phi_v^T M_{ib_2} & \cdots & \Phi_v^T M_{ii}\Phi_{cb_n} + \Phi_v^T M_{ib_n} & \Phi_v^T M_{ii}\Phi_v \end{bmatrix} \tag{3.3.21}$$

### Blade Number $j$

Let the displacement vector of the interior degrees of freedom of the blade number $j$ be $\mathbf{u}_i^{bl_j}$, and that for boundary nodes is $\mathbf{u}_{b_j}$ (Figure 3.3.1). Let the mass and stiffness matrix of the blade number $j$ be

$$M^{bl_j} = \begin{bmatrix} M_{b_j b_j}^{bl} & M_{b_j i}^{bl} \\ M_{ib_j}^{bl} & M_{ii}^{bl_j} \end{bmatrix} \tag{3.3.22}$$

$$K^{bl_j} = \begin{bmatrix} K_{b_j b_j}^{bl} & K_{b_j i}^{bl} \\ K_{ib_j}^{bl} & K_{ii}^{bl_j} \end{bmatrix} \tag{3.3.23}$$

Free vibration of the blade number $j$ will be described by

$$M^{bl_j}\ddot{\mathbf{u}}_{bl_j} + K^{bl_j}\mathbf{u}_{bl_j} = \mathbf{0} \tag{3.3.24}$$

$$\mathbf{u}_{bl_j}^T = \begin{bmatrix} \mathbf{u}_{b_j}^T & \mathbf{u}_i^{bl_j^T} \end{bmatrix} \tag{3.3.25}$$

### Constrained Static Modes for Blade Number $j$

Let the force vector of the interior degrees of freedom of the blade be $\mathbf{f}_i^{bl}$, and that for boundary nodes be $\mathbf{f}_{b_j}^{bl}$. Then, static equilibrium equations will be given by

$$
\begin{bmatrix} K_{b_jb_j}^{bl} & K_{b_ji}^{bl} \\[2mm] K_{ib_j}^{bl} & K_{ii}^{bl_j} \end{bmatrix}
\begin{bmatrix} \mathbf{u}_{b_j} \\[2mm] \mathbf{u}_i^{bl_j} \end{bmatrix}
=
\begin{bmatrix} \mathbf{f}_{b_j}^{bl} \\[2mm] \mathbf{f}_i^{bl} \end{bmatrix}
\tag{3.3.26}
$$

With $\mathbf{f}_i^{bl} = 0$,

$$
K_{ib_j}^{bl}\,\mathbf{u}_{b_j} + K_{ii}^{bl_j}\,\mathbf{u}_i^{bl_j} = 0
\tag{3.3.27}
$$

Therefore,

$$
\mathbf{u}_i^{bl_j} = \Phi_{cb_j}^{bl}\,\mathbf{u}_{b_j}
\tag{3.3.28}
$$

where

$$
\Phi_{cb_j}^{bl} = -K_{ii}^{bl_j^{-1}}\,K_{ib_j}^{bl}
\tag{3.3.29}
$$

Let there be unit displacement for each element of $\mathbf{u}_{b_j}$. In this case, a constrained displacement vector can be obtained using Equation (3.3.28) as follows:

$$
\begin{bmatrix} \mathbf{u}_{b_j} \\[2mm] \mathbf{u}_i^{bl_j} \end{bmatrix}
=
\begin{bmatrix} I_{b_j} \\[2mm] \Phi_{cb_j}^{bl} \end{bmatrix}
\tag{3.3.30}
$$

## Constrained Vibratory Modes for Blade Number $j$

Let displacements at blade boundary nodes be zero, that is,

$$
\mathbf{u}_{b_j} = 0
\tag{3.3.31}
$$

In this case, Equations (3.3.22)–(3.3.25) lead to following differential equations of motion

$$
M_{ii}^{bl_j}\,\ddot{\mathbf{u}}_i^{bl_j} + K_{ii}^{bl_j}\,\mathbf{u}_i^{bl_j} = \mathbf{0}
\tag{3.3.32}
$$

The corresponding modal matrix is written as

$$
K_{ii}^{bl_j}\,\Phi_v^{bl_j} = M_{ii}^{bl_j}\,\Phi_v^{bl_j}\,\Lambda_d^{bl_j}
\tag{3.3.33}
$$

where

$$
\Phi_v^{bl_j^{T}}\,K_{ii}^{bl_j}\,\Phi_v^{bl_j} = \Lambda_d^{bl_j} \quad \text{and} \quad \Phi_v^{bl_j^{T}}\,M_{ii}^{bl_j}\,\Phi_v^{bl_j} = I
\tag{3.3.34}
$$

Here, $\Lambda_d^{bl_j}$ is a diagonal matrix containing eigenvalues, and the matrix $\Phi_v^{bl_j}$ is composed of eigenvectors.

### Model Reduction for Blade Number j

The displacement vector of blade number j, Equation (3.3.25), is approximated as

$$\mathbf{u}_{bl_j} = \mathbf{\Psi}_{bl_j}\,\mathbf{\alpha}_{bl_j} \tag{3.3.35}$$

where

$$\mathbf{\alpha}_{bl_j}^T = \begin{bmatrix} \mathbf{\alpha}_{b_j}^{bl^T} & \mathbf{\alpha}_i^{bl_j^T} \end{bmatrix} \tag{3.3.36}$$

and

$$\mathbf{\Psi}_{bl_j} = \begin{bmatrix} I_{b_j} & 0 \\ \Phi_{cb_j}^{bl} & \Phi_v^{bl_j} \end{bmatrix} \tag{3.3.37}$$

Let the dimension of $\mathbf{\alpha}_{b_j}^{bl}$ be $n_{b_j}$ and the number of vibratory modes in matrix $\Phi_v^{bl_j}$ be $nv_j^{bl}$. Then, the dimension of $\mathbf{\alpha}_{bl_j}$ will be $nv_j^{bl} + n_{b_j}$. Substituting Equation (3.3.35) into Equation (3.3.24),

$$\hat{M}^{bl_j}\,\ddot{\mathbf{\alpha}}_{bl_j} + \hat{K}^{bl_j}\,\mathbf{\alpha}_{bl_j} = 0 \tag{3.3.38}$$

where

$$\hat{M}^{bl_j} = \mathbf{\Psi}_{bl_j}^T\,M^{bl_j}\,\mathbf{\Psi}_{bl_j} = \begin{bmatrix} M_{b_j b_j}^{bl} + \Phi_{cb_j}^{bl^T} M_{ib_j}^{bl} & M_{b_j i}^{bl}\Phi_v^{bl_j} + \\ + M_{b_j i}^{bl}\Phi_{cb_j}^{bl} + \Phi_{cb_j}^{bl^T} M_{ii}^{bl_j}\Phi_{cb_j}^{bl} & \Phi_{cb_j}^{bl^T} M_{ii}^{bl_j}\Phi_v^{bl_j} \\ & \\ \Phi_v^{bl_j^T} M_{ib_j}^{bl} + \Phi_v^{bl_j^T} M_{ii}^{bl_j}\Phi_{cb_j}^{bl} & \Phi_v^{bl_j^T} M_{ii}^{bl_j}\Phi_v^{bl_j} \end{bmatrix} \tag{3.3.39}$$

and

$$\hat{K}^{bl_j} = \mathbf{\Psi}_{bl_j}^T\,K^{bl_j}\,\mathbf{\Psi}_{bl_j} = \begin{bmatrix} K_{b_j b_j}^{bl} + \Phi_{cb_j}^{bl^T} K_{ib_j}^{bl} & 0 \\ 0 & \Phi_v^{bl_j^T} K_{ii}^{bl_j}\Phi_v^{bl_j} \end{bmatrix} \tag{3.3.40}$$

### Reduced-Order Model for Bladed Disk

Reduced-order model for the complete bladed disk is obtained by combining reduced-order models of disk, Equation (3.3.17), and all blades, Equation (3.3.38), as follows:

$$\hat{M}_{bd}\,\ddot{\mathbf{\alpha}}_{bd} + \hat{K}_{bd}\,\mathbf{\alpha}_{bd} = 0 \tag{3.3.41}$$

where

$$\alpha_{bd}^T = \begin{bmatrix} \alpha_d^T & \alpha_{bl_1}^T & \alpha_{bl_2}^T & \cdots & \alpha_{bl_n}^T \end{bmatrix} \tag{3.3.42}$$

where vector $\alpha_d$ is defined by Equation (3.3.15) and vector $\alpha_{bl_j}$ by Equation (3.3.36). Further,

$$\hat{K}_{bd} = \begin{bmatrix} \hat{K}^d & & & & \\ & \hat{K}^{bl_1} & & & \\ & & \hat{K}^{bl_2} & & \\ & & & \ddots & \\ & & & & \hat{K}^{bl_n} \end{bmatrix} \tag{3.3.43}$$

and

$$\hat{M}_{bd} = \begin{bmatrix} \hat{M}^d & & & & \\ & \hat{M}^{bl_1} & & & \\ & & \hat{M}^{bl_2} & & \\ & & & \ddots & \\ & & & & \hat{M}^{bl_n} \end{bmatrix} \tag{3.3.44}$$

It is required that

$$\alpha_{b_1} = \mathbf{u}_{b_1}, \ \alpha_{b_2} = \mathbf{u}_{b_2}, \ldots, \alpha_{b_n} = \mathbf{u}_{b_n} \tag{3.3.45a}$$

$$\alpha_{b_1}^{bl} = \mathbf{u}_{b_1}, \alpha_{b_2}^{bl} = \mathbf{u}_{b_2}, \ldots, \alpha_{b_n}^{bl} = \mathbf{u}_{b_n} \tag{3.3.45b}$$

It should be noted that $\alpha_{b_1}, \alpha_{b_2}, \ldots, \alpha_{b_n}$ are parts of the vector $\alpha_d$, Equation (3.3.15), and $\alpha_{b_j}^{bl}$ is a part of vector $\alpha_{bl_j}$, Equation (3.3.36). To impose boundary conditions, a matrix S is defined as follows:

$$\alpha_{bd} = S \, \mathbf{u}_{bd} \tag{3.3.46}$$

where

$$\mathbf{u}_{bd}^T = \begin{bmatrix} \mathbf{u}_{b_1}^T & \mathbf{u}_{b_2}^T & \cdots & \mathbf{u}_{b_n}^T & \alpha_i^T & \alpha_i^{bl_1^T} & \cdots & \alpha_i^{bl_n^T} \end{bmatrix} \tag{3.3.47}$$

and

$$S = \begin{bmatrix} I & 0 & \cdots & 0 & 0 & 0 & \cdots & 0 \\ 0 & I & \cdots & 0 & 0 & 0 & \cdots & 0 \\ \cdots & \cdots & \cdots & \cdots & \cdots & \cdots & \cdots & \cdots \\ 0 & 0 & \cdots & I & 0 & 0 & \cdots & 0 \\ 0 & 0 & \cdots & 0 & I & 0 & \cdots & 0 \\ I & 0 & \cdots & 0 & 0 & 0 & \cdots & 0 \\ 0 & 0 & \cdots & 0 & 0 & I & \cdots & 0 \\ 0 & I & \cdots & 0 & 0 & 0 & \cdots & 0 \\ 0 & 0 & \cdots & 0 & 0 & 0 & \cdots & 0 \\ \vdots & \vdots & \cdots & \vdots & \vdots & \vdots & \cdots & \vdots \\ 0 & 0 & \cdots & I & 0 & 0 & \cdots & 0 \\ 0 & 0 & \cdots & 0 & 0 & 0 & \cdots & I \end{bmatrix} \tag{3.3.48}$$

Substituting Equation (3.3.46) into Equation (3.3.41) and premultiplying both sides by $S^T$,

$$M_{rbd}\ddot{\mathbf{u}}_{bd} + K_{rbd}\mathbf{u}_{bd} = \mathbf{0} \tag{3.3.49}$$

where

$$M_{rbd} = S^T \hat{M}_{bd} S \tag{3.3.50}$$

$$K_{rbd} = S^T \hat{K}_{bd} S \tag{3.3.51}$$

Equations (3.3.49)–(3.3.51) represents the reduced-order model obtained through the CMS approach.

## 3.4    Reduced-Order Model: Modified Modal Domain Analysis (MMDA)

This section deals with the geometric mistuning that refers to variations in geometries of blades, for example, in an integrally bladed rotor (IBR). Geometries of blades of an IBR are measured by a coordinate measurement machine (CMM). *The distinguishing feature of geometric mistuning is that it leads to simultaneous and dependent perturbations in both mass and stiffness matrices.* Sinha et al. (2008) have shown that vibratory parameters of an individual blade can be extracted from CMM data using POD (proper orthogonal decomposition) analyses. The objective is to develop a reduced-order model of a bladed disk with geometric mistuning. First, the SNM or modal domain analysis (MDA) (Section 3.2) was applied to develop the reduced-order model. However, it did not succeed in the presence of simultaneous perturbations in mass and stiffness matrices due to geometric mistuning. This frustrated many researchers because it was difficult to reconcile with the expectation that any mistuned mode must be a linear combination of a subset of tuned modes. After some frustrating moments, the author of this monograph realized that the failure of MDA in modeling geometric mistuning is not because any mistuned mode cannot be a linear combination of a subset of tuned modes, but because of insufficient number of tuned modes. As the MDA model with an arbitrarily large number of tuned modes cannot be called reduced-order, the MDA method has been modified by Sinha (2009) to include tuned modes with blades having geometries perturbed along important POD features (Sinha et al., 2008) as basis functions. The resulting method, Modified Modal Domain Analysis (MMDA), has yielded a high-fidelity reduced-order model. This result is a major breakthrough in the jet engine vibration research. An important aspect of MMDA is that the reduced-order model can be obtained from finite element-sector analyses only.

## 3.4.1    MMDA Algorithm

A mistuned bladed-disk assembly or a bladed rotor can be described by

$$M\ddot{\mathbf{x}} + C\dot{\mathbf{x}} + K\mathbf{x} = \mathbf{f}(t) \tag{3.4.1}$$

where $M$, $K$, and $C$ are mass, stiffness, and damping matrices, respectively. The external excitation vector is represented by $\mathbf{f}$. Define

$$M = M_t + \delta M \tag{3.4.2}$$

and

$$K = K_t + \delta K \tag{3.4.3}$$

where $M_t$ and $K_t$ are mass and stiffness matrices of a perfectly tuned system, respectively. Matrices $\delta M$ and $\delta K$ are deviations in mass and stiffness matrices due to mistuning.

The geometry of a blade can be described by

$$\mathbf{w}(\mathbf{p}) = \bar{\mathbf{w}}(\mathbf{p}) + \delta\mathbf{w}(\mathbf{p}) \tag{3.4.4}$$

where $\mathbf{w}(\mathbf{p})$ is coordinate of the blade surface location described by the vector $\mathbf{p}$. The mean geometry of blade is denoted by $\bar{\mathbf{w}}(\mathbf{p})$, and $\delta\mathbf{w}(\mathbf{p})$ is the random deviation in the geometry of blade. Let the covariance matrix of random $\delta\mathbf{w}(\mathbf{p})$ be $C(\mathbf{p}_i, \mathbf{p}_\ell)$. Using Karhunen-Loeve expansion (Ghanem and Spanos, 1990)

$$\delta\mathbf{w}(\mathbf{p}) = \sum_\ell \xi_\ell \sqrt{\lambda_\ell} \mathbf{u}_\ell \tag{3.4.5}$$

where $\lambda_\ell$ and $\mathbf{u}_\ell$ are eigenvalue and eigenvector of the covariance matrix $C(\mathbf{p}_i, \mathbf{p}_\ell)$, and $\xi_\ell$ = uncorrelated random variables with zero mean and unity standard deviation.

In other words,

$$E(\xi_\ell) = 0; \quad E(\xi_\ell^2) = 1 \tag{3.4.6}$$

and

$$E(\xi_\ell \xi_i) = 0 \quad \text{when} \quad \ell \neq i \tag{3.4.7}$$

The Karhunen-Loeve expansion is also called Proper Othogonal Decomposition (POD). It should be noted that the POD expansion (Equation 3.4.5) is based on the second-order statistics, which is sufficient for a Gaussian distribution. It does not require perturbations in the geometry to be small.

The vector $\mathbf{u}_\ell$ is also called $\ell th$ POD feature. Typically, values of $\lambda_\ell$ are almost equal to zero except for few POD features. Hence, considering only $np$ largest values of $\lambda_\ell$, Equation (3.4.5) can be approximated as

$$\delta \mathbf{w}(\mathbf{p}) = \sum_{\ell=1}^{np} \xi_\ell \sqrt{\lambda_\ell} \, \mathbf{u}_\ell \tag{3.4.8}$$

The computation of eigenvalues and eigenvectors of the covariance matrix is not efficient. Therefore, POD features and $\sqrt{\lambda_\ell}$ are computed by singular value decomposition, which is presented in Section 3.5.2.

Now, the transformation (3.2.3) is modified as follows:

$$\mathbf{x} = \boldsymbol{\Phi} \mathbf{y} \tag{3.4.9}$$

where

$$\boldsymbol{\Phi} = [\boldsymbol{\Phi}_0 \quad \boldsymbol{\Phi}_1 \quad \cdot \quad \cdot \quad \cdot \quad \boldsymbol{\Phi}_{np}] \tag{3.4.10}$$

$\boldsymbol{\Phi}_0$ : $r$ tuned modes of the system with blades having the mean geometry $\quad$ (3.4.11)

$\boldsymbol{\Phi}_\ell$ : $r$ tuned modes of system with blades having perturbed
geometry along $\ell th$ POD feature, $\ell = 1,....,np$ $\qquad$ (3.4.12)

In this case, the dimension of the reduced order system will be $r(np+1)$. The geometry of the blade that is perturbed along $\ell^{th}$ POD feature is described by

$$\mathbf{w}(\mathbf{p}) = \overline{\mathbf{w}}(\mathbf{p}) + \xi_\ell \sqrt{\lambda_\ell} \, \mathbf{u}_\ell \tag{3.4.13}$$

Hence, the matrix $\boldsymbol{\Phi}_\ell$ is composed of tuned modes of the rotor with all blades having the geometry defined by Equation (3.4.13). Substituting Equation (3.4.9) into Equation (3.4.1) and premultiplying by $\boldsymbol{\Phi}^H$, the following reduced-order system is obtained:

$$M_r \ddot{\mathbf{y}} + C_r \dot{\mathbf{y}} + K_r \mathbf{y} = \boldsymbol{\Phi}^H \mathbf{f}(t) \tag{3.4.14}$$

where

$$K_r = \boldsymbol{\Phi}^H K \boldsymbol{\Phi} = \boldsymbol{\Phi}^H K_t \boldsymbol{\Phi} + \boldsymbol{\Phi}^H \, \delta K \boldsymbol{\Phi} \tag{3.4.15}$$

$$M_r = \Phi^H M \Phi = \Phi^H M_t \Phi + \Phi^H \delta M \Phi \tag{3.4.16}$$

$$C_r = \Phi^H C \Phi \tag{3.4.17}$$

The natural frequencies and mode shapes of a mistuned system can be obtained by solving the eigenvalue problem corresponding to the reduced-order system:

$$K_r \psi = \omega_r^2 M_r \psi \tag{3.4.18}$$

The reduced-order system is valid when the natural frequencies and mode shapes predicted by (3.4.18) are same as those predicted by the full-order system:

$$K \varphi = \omega^2 M \varphi \tag{3.4.19}$$

It should be noted that the equivalent representation of (Equation 3.4.10) can also be written as $\Phi_{eq} = [\Phi_0 \quad (\Phi_1 - \Phi_0) \quad \cdots \quad (\Phi_{np} - \Phi_0)]$. Therefore, the use of $\Phi_\ell$; $\ell = 1, ...., np$, as basis functions provides information about the effects of the change in geometry along the $\ell th$ POD feature with respect to $\Phi_0$. Qualitatively, $\Phi_\ell - \Phi_0$ is almost orthogonal to $\Phi_0$, that is, effects of geometry change are almost orthogonal to $\Phi_0$ and as a result $\Phi_0$ alone is not able to capture the effects of geometric mistuning.

To obtain reduced-order mass and stiffness matrices, $\Phi_i^H K_t \Phi_j, \Phi_i^H M_t \Phi_j, \Phi_i^H \delta K \Phi_j$, and $\Phi_i^H \delta M \Phi_j$ are to be calculated for $i = 0,1,2,...,np$, and $j = i, i+1,...,np$. Here, only algorithms to compute $\Phi_i^H M_t \Phi_j$ and $\Phi_i^H \delta M \Phi_j$ are presented (Sinha and Bhartiya, 2010). The procedures to compute $\Phi_i^H K_t \Phi_j$ and $\Phi_i^H \delta K \Phi_j$ are similar.

## Computation of $\Phi_i^H M_t \Phi_j$ Using ANSYS Sector Analyses

Let $\varphi_{i,j,p}$ represent nodal displacements associated with the tuned mode for nominal or geometry perturbed along a POD feature ($i = 0,1,2,...,np$), sector $j$ and the interblade phase angle $= 2\pi p / n$. Note that

$$\varphi_{i,j,p} = e^{\iota(j-1)\psi_p} \varphi_{i,1,p}; \quad \iota = \sqrt{-1}; \tag{3.4.20}$$

where

$$\psi_p = \frac{2\pi p}{n} \tag{3.4.21}$$

The $(\ell+1, \rho+1)$ element of $\Phi_i^H M_t \Phi_j$ is as follows:

$$\varphi_{i,1,\ell}^H M_{t\rho}^c \varphi_{j,1,\rho} + \varphi_{i,2,\ell}^H M_{t\rho}^c \varphi_{j,2,\rho} + \cdots + \varphi_{i,n,\ell}^H M_{t\rho}^c \varphi_{j,n,\rho}$$

$$= (1 + e^{\iota(\psi_\rho - \psi_\ell)} + \cdots + e^{\iota(n-1)(\psi_\rho - \psi_\ell)}) \varphi_{i,1,\ell}^H M_{t\rho}^c \varphi_{j,1,\rho} = \chi \varphi_{i,1,\ell}^H M_{t\rho}^c \varphi_{j,1,\rho} \tag{3.4.22}$$

where

$$M_{t\rho}^c = M_{t,1,1} + e^{\iota\psi_\rho} M_{t,1,2} + \cdots + e^{\iota(n-1)\psi_\rho} M_{t,1,n} \tag{3.4.23}$$

and

$$\chi = \frac{1 - e^{\imath n (\psi_\rho - \psi_\ell)}}{1 - e^{\imath (\psi_\rho - \psi_\ell)}}$$

(3.4.24)

Let $\chi_R$ and $\chi_I$ be real and imaginary parts of $\chi$, respectively. With $M_{t\rho}^{cR}$ and $M_{t\rho}^{cI}$ being real and imaginary parts of $M_{t\rho}^{c}$,

$$\chi M_{t\rho}^{c} = M_{t\rho}^{mcR} + \imath M_{t\rho}^{mcI}; \quad \imath = \sqrt{-1}$$

(3.4.25)

where

$$M_{t\rho}^{mcR} = (\chi_R M_{t\rho}^{cR} - \chi_I M_{t\rho}^{cI}) \quad \text{and} \quad M_{t\rho}^{mcI} = (\chi_I M_{t\rho}^{cR} + \chi_R M_{t\rho}^{cI})$$

(3.4.26)

Let

$$\varphi_{j,1,\rho} = \varphi_{j,1,\rho}^{R} + \imath \varphi_{j,1,\rho}^{I}$$

(3.4.27)

From Equation (3.4.22), the $(\ell + 1, \rho + 1)$ element of $\Phi_i^H M_t \Phi_j$ is further expressed as follows:

$$\chi \varphi_{i,1,\ell}^H M_{t\rho}^{c} \varphi_{j,1,\rho} = \varphi_{i,1,\ell}^H M_{t\rho}^{mc} \varphi_{j,1,\rho} = [\varphi_{i,1,\ell}^{R'} \quad \varphi_{i,1,\ell}^{I'}] \begin{bmatrix} M_{t\rho}^{mcR} & -M_{t\rho}^{mcI} \\ M_{t\rho}^{mcI} & M_{t\rho}^{mcR} \end{bmatrix} \begin{bmatrix} \varphi_{j,1,\rho}^{R} \\ \varphi_{j,1,\rho}^{I} \end{bmatrix}$$
$$+ \imath [-\varphi_{i,1,\ell}^{I'} \quad \varphi_{i,1,\ell}^{R'}] \begin{bmatrix} M_{t\rho}^{mcR} & -M_{t\rho}^{mcI} \\ M_{t\rho}^{mcI} & M_{t\rho}^{mcR} \end{bmatrix} \begin{bmatrix} \varphi_{j,1,\rho}^{R} \\ \varphi_{j,1,\rho}^{I} \end{bmatrix}$$

(3.4.28)

From Equation (3.4.26),

$$\begin{bmatrix} M_{t\rho}^{mcR} & -M_{t\rho}^{mcI} \\ M_{t\rho}^{mcI} & M_{t\rho}^{mcR} \end{bmatrix} = \chi_R \begin{bmatrix} M_{t\rho}^{cR} & -M_{t\rho}^{cI} \\ M_{t\rho}^{cI} & M_{t\rho}^{cR} \end{bmatrix} + \chi_I \begin{bmatrix} -M_{t\rho}^{cI} & -M_{t\rho}^{cR} \\ M_{t\rho}^{cR} & -M_{t\rho}^{cI} \end{bmatrix}$$

(3.4.29)

### Computation of $\Phi_i^H \delta M \Phi_j$ Using ANSYS Sector Analyses

The perturbation in the mass matrix, $\delta M$, has the block diagonal form. Therefore,

$$\Phi_i^H \delta M \Phi_j = \sum_{\ell=1}^{n} \Phi_i^H \delta M_\ell^b \Phi_j$$

(3.4.30)

where

$$\delta M_\ell^b = \begin{bmatrix} 0 & \cdots & 0 & \cdots & 0 \\ \cdots & \ddots & \cdots & \cdots & \cdots \\ 0 & \cdots & \delta M_\ell & \cdots & 0 \\ \cdots & \cdots & \cdots & \ddots & \cdots \\ 0 & \cdots & 0 & \cdots & 0 \end{bmatrix}$$

(3.4.31)

The $(\nu+1, \rho+1)$ element of $\Phi_i^H \delta M_\ell^b \Phi_j$ in Equation (3.4.30) is as follows:

$$\varphi_{i,1,\nu}^H \delta M_\ell e^{\iota\ell(\psi_\rho-\psi_\nu)} \varphi_{j,1,\rho} = [\varphi_{i,1,\nu}^{R'} \quad \varphi_{i,1,\nu}^{I'}] \begin{bmatrix} \delta M_{\ell c} & -\delta M_{\ell s} \\ \delta M_{\ell s} & \delta M_{\ell c} \end{bmatrix} \begin{bmatrix} \varphi_{j,1,\rho}^R \\ \varphi_{j,1,\rho}^I \end{bmatrix}$$
$$+ \iota[-\varphi_{i,1,\nu}^{I'} \quad \varphi_{i,1,\nu}^{R'}] \begin{bmatrix} \delta M_{\ell c} & -\delta M_{\ell s} \\ \delta M_{\ell s} & \delta M_{\ell c} \end{bmatrix} \begin{bmatrix} \varphi_{j,1,\rho}^R \\ \varphi_{j,1,\rho}^I \end{bmatrix} \qquad (3.4.32)$$

where

$$\delta M_{\ell c} = \delta M_\ell \cos(\ell(\psi_\rho - \psi_\nu)) \quad \text{and} \quad \delta M_{\ell s} = \delta M_\ell \sin(\ell(\psi_\rho - \psi_\nu)) \qquad (3.4.33a, b)$$

### Connection with ANSYS Sector Analysis

The ANSYS code only works with real numbers (ANSYS, 2012). It constructs two identical sectors, one for real part and another for imaginary part of the displacement (Section 3.1.1). As a result, only one interblade phase angle for a repeated eigenvalue is considered. Further, the mass and stiffness matrices for a double sector are as follows:

$$\begin{bmatrix} M_{t\rho}^{cR} & -M_{t\rho}^{cI} \\ M_{t\rho}^{cI} & M_{t\rho}^{cR} \end{bmatrix} \text{and} \begin{bmatrix} K_{t\rho}^{cR} & -K_{t\rho}^{cI} \\ K_{t\rho}^{cI} & K_{t\rho}^{cR} \end{bmatrix} \qquad (3.4.34a, b)$$

Therefore, $M_{t\rho}^{cR}$, $M_{t\rho}^{cI}$ in Equation (3.4.29) can be obtained directly from ANSYS mass matrices. Similarly, $K_{t\rho}^{cR}$, $K_{t\rho}^{cI}$ can be obtained directly from ANSYS stiffness matrices. Furthermore, eigenvectors of the double sector have exactly the form $[\varphi_{i,1,\ell}^{R'} \quad \varphi_{i,1,\ell}^{I'}]$ and $[-\varphi_{i,1,\ell}^{I'} \quad \varphi_{i,1,\ell}^{R'}]$, which appear in Equations (3.4.28) and (3.4.32).

## 3.4.2 Alternate MMDA Bases Vectors Directly from Mistuned Sectors

In MMDA the true mode shapes of a mistuned bladed-disk assembly are approximated by a linear combination of mode shapes of "average" tuned geometry and tuned geometries of sectors with blades perturbed along the POD features as given by Equation (3.4.10). The modes for the geometries from the POD analysis are used because POD analysis provides independent vectors for perturbations in geometries, and by using only dominant POD features to form the bases of geometric perturbations; a minimal set of mode shapes is obtained to form the solution bases. The idea behind POD analysis is to obtain independent perturbation vectors and because the perturbation in actual sector can be represented as a linear combination of the POD vectors using Karhunen–Loeve (KL) expansion (Sinha et al., 2008), it is proposed (Bhartiya and Sinha, 2013b) that alternatively the actual mistuned sectors may themselves be used to form suitable bases, that is,

$$\Phi = [\Phi_0 \quad \Phi_{a1} \quad \Phi_{a2} \quad . \quad . \quad \Phi_{an_q}] \qquad (3.4.35)$$

where $\Phi_{a\ell}$: mode shapes of the tuned bladed disk with each sector represented by the sector number $\ell$ of the actual mistuned bladed disk, and $n_q$ is the required number of actual sectors.

### 3.4.3     MMDA for Bladed Rotor with Rogue Blades

Another problem that arises in mistuning is that of the rogue blade, that is, a blade having geometry significantly different from the average geometry, which can be caused by foreign object damage (FOD) or blade-tip blending to remove blade corrosion. In such cases the mode shape of the rogue blade is significantly different from the "average" mode shape and the techniques discussed so far may fail to provide accurate results. But as it has been observed from the results of alternative bases, the mode shapes of actual blades can be used to form the bases. Same idea can be extended in case of extremely large mistuning and the mode shapes from the cyclic analysis of the sector with rogue blade, $\Phi_{rogue}$, can also be included in bases to consider the impact of extremely large mistuning. Hence in the presence of extremely large mistuning the following modification to the bases in MMDA algorithm is suggested (Bhartiya and Sinha, 2013b):

$$\Phi = [\Phi_0 \quad \Phi_1 \quad \Phi_2 \quad . \quad . \quad \Phi_{np} \quad \Phi_{rogue}] \tag{3.4.36}$$

Equation (3.4.36) is similar to Equation (3.4.10) where $\Phi_0$ are tuned modes of the system with blades having the mean geometry and $\Phi_\ell$ ($\ell = 1, ...., np$) are tuned modes of the system with blades having geometry perturbed along $\ell^{th}$ POD feature (the rogue blade is not included in the POD analysis). $\Phi_{rogue}$ is composed of the tuned modes from the cyclic analysis of the sector with the rogue blade. The explicit inclusion of the mode shapes from the rogue blade in the bases is done to account for the large changes in the mode shapes of the rogue blade.

## 3.5     Numerical Examples: Natural Frequencies, Mode Shapes, and Forced Response

### 3.5.1     Academic Rotor

An ANSYS model of an academic rotor (Figure 3.5.1) has been constructed. This model has twenty-four blades. For each blade, the finite element grid has six circumferential locations, numbered 1 to 6 in Figure 3.5.2, and nodes are only on the front and back surfaces in $xy$ plane, with $z$-coordinates being either $+q$ or $-q$ inches, where $2q$ is the thickness of each tuned blade. Therefore, the thickness of the blade is changed at any circumferential location by multiplying the z-coordinate of each node at that circumferential location by the same factor.

To have each blade uniform but a different thickness, all nodal z-coordinates of a blade are multiplied by

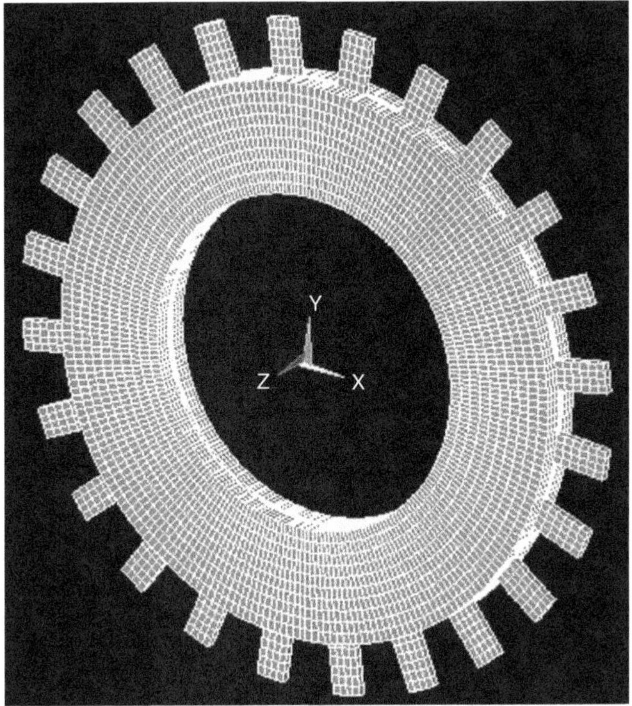



**Figure 3.5.1.**   An academic rotor with twenty-four blades (Bhartiya and Sinha, 2011).

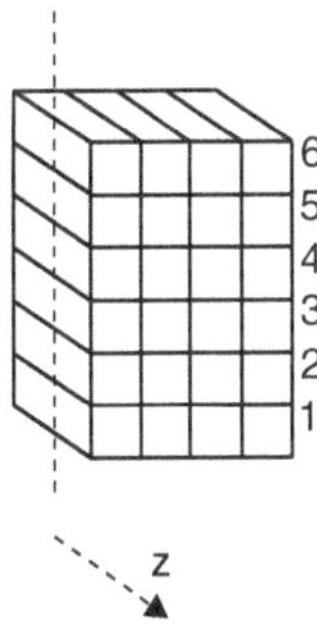



**Figure 3.5.2.**   Discretization of a blade by finite elements (Sinha, 2009).

$$1+\xi_{1i}; \quad i = 1, 2, \dots\dots, 24 \tag{3.5.1}$$

where $\xi_1$ is the random variable represensing the only POD feature, and $\xi_{1i}$ is its value for blade number $i$. Based on the multiplying factor (Equation 3.5.1), a mistuning pattern (#1) with mistuned uniform blade thicknesses is created (see Figure 3.5.3).

Figure 3.5.4 shows the natural frequencies of the nominal bladed-disk assembly for different families of modes. As observed from the figure, two types of regions exist: (1) regions with isolated family of modes in a narrow frequency band where

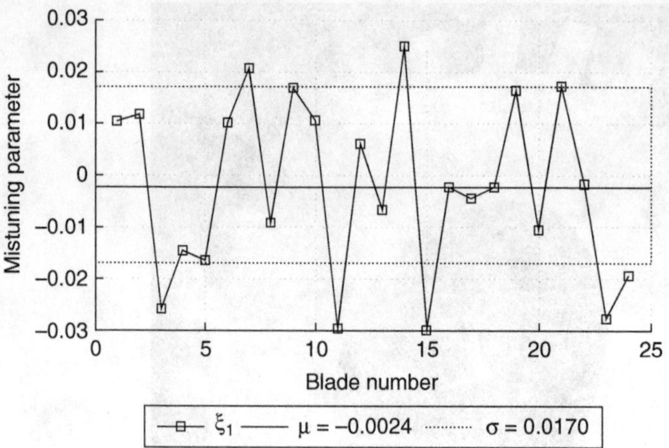

**Figure 3.5.3.**    Mistuning pattern for blade thicknesses (Bhartiya and Sinha, 2011).

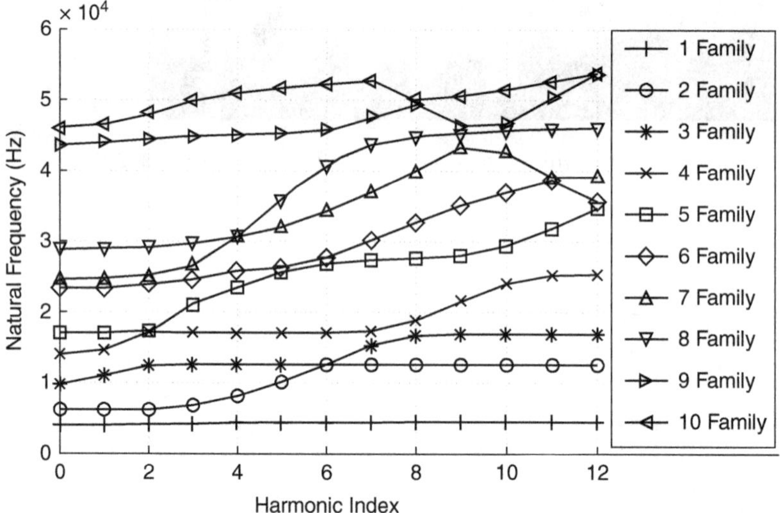

**Figure 3.5.4.**    Natural frequencies versus harmonic index (Bhartiya and Sinha, 2011).

the primary energy is stored in the blades, for example, family 1; and (2) regions with overlapping families spanning a larger frequency bands where the primary energy is stored in the disk, for example, families 4, 5, and 6. From a frequency mistuning point of view these two regions are different in the sense that for isolated families only a single-blade mode shape is present in the region, and natural frequency of the mistuned blades for that mode can be used to calculate the equivalent Young's moduli of elasticity of blades for frequency mistuning. But in the other region where multiple families overlap, multiple definitions of equivalent frequency mistuning exist depending upon the family of modes used to calculate

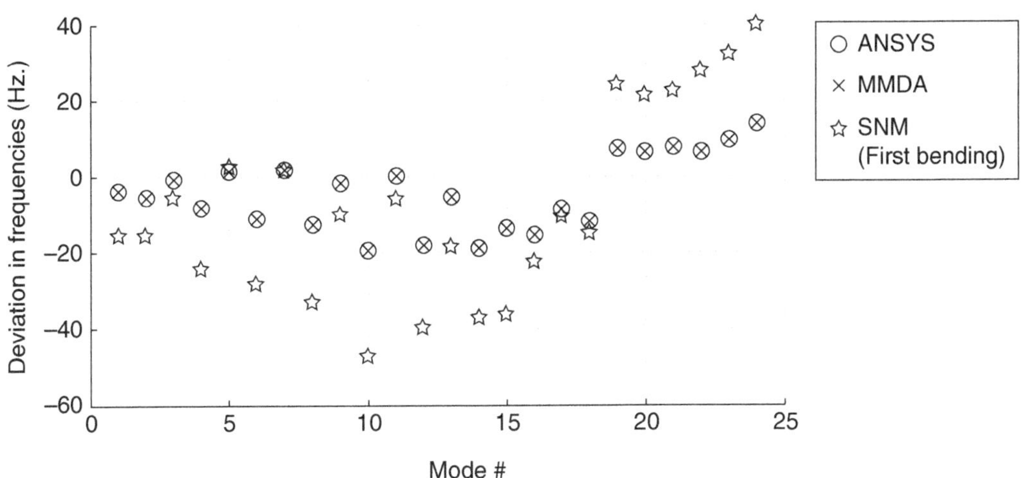


**Figure 3.5.5.**   Deviations in frequencies estimated using reduced-order models (MMDA and SNM) for the first bending family (Bhartiya and Sinha, 2011).

the equivalent Young's moduli of the blades. Both the cases for the frequency mistuning have been considered in this study.

Figure 3.5.5 shows deviations in the first twenty-four natural frequencies estimated using MMDA, SNM, and ANSYS (full 360-degree rotor) analysis. As observed from the figure, SNM is unable to capture the deviations in natural frequencies due to geometric mistuning with standard deviation equal to 1.7 percent.

Next the mode shapes from the reduced-order models (MMDA and SNM) are compared with the mode shapes from the full rotor ANSYS analysis using modal assurance criterion (MAC) (Allemang, 2003). MAC values for the mode shapes estimated using reduced-order models are plotted in Figure 3.5.6. The values closer to 1 on the diagonal suggest that the mode shapes estimated from the reduced-order model are identical to the reference mode shapes (mode shapes from full-rotor ANSYS analysis), whereas the values closer to 0 on the diagonal suggest that the estimated mode shapes from the reduced-order model are orthogonal to the reference mode shapes.

The observation of MAC values for mode shapes estimated using MMDA suggests that MMDA is able to capture the mode shapes exactly. On the other hand, MAC values for the modes estimated using SNM suggest that the technique is able to capture mode shapes for modes 1–12, but shows large errors in estimated mode shapes for modes 13–24.

A closer look at the mode shapes of the bladed-disk assembly shows that the first twelve modes do not show significant mode localization (for example, mode 5 in Figure 3.5.7a) and are hence similar to the modes of the nominal tuned bladed-disk assembly. For this reason, nominal mode approximation is sufficient to estimate the first twelve mode shapes of the bladed-disk assembly. However, modes 13–24

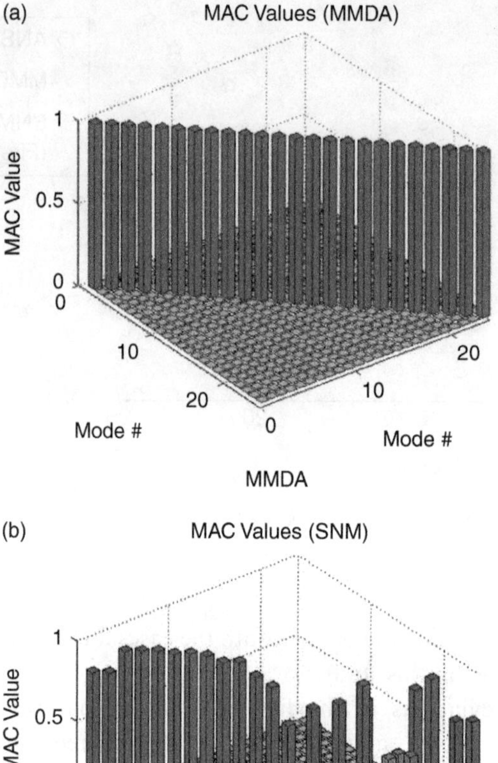



**Figure 3.5.6.** MAC values for the first twenty-four modes calculated using reduced-order models (MMDA and SNM) for the first bending family (Bhartiya and Sinha, 2011).

show significant mode localization (for example, mode 19 in Figure 3.5.7b) and are different from the mode shapes of the nominal tuned bladed-disk assembly. In this case, the nominal mode approximation of the mistuned modes is not sufficient and an additional set of nonnominal modes is required to form a suitable basis for the mistuned mode shapes.

Similar analysis for comparison between SNM and MMDA is also performed for frequency band near 22 kHz. Figure 3.5.8 shows the deviations in frequencies estimated using MMDA, SNM, and full-rotor ANSYS analysis. As observed from the figure, MMDA is able to capture the effects of geometric mistuning exactly whereas errors are observed in the frequency deviation estimates from

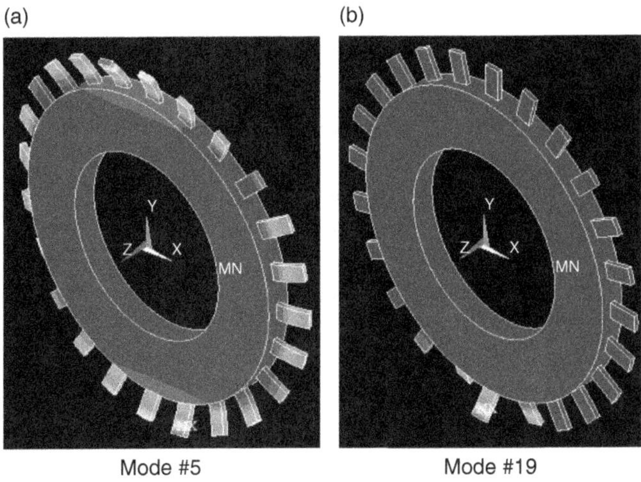



**Figure 3.5.7.**   Mode shapes of mistuned bladed-disk assembly (Bhartiya and Sinha, 2011).

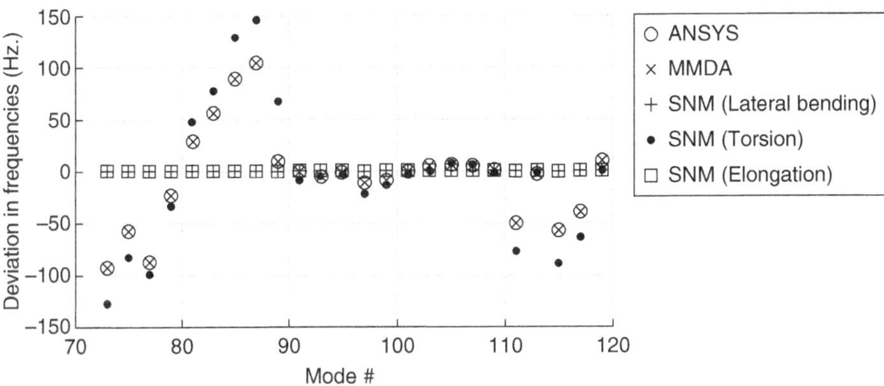



**Figure 3.5.8.**   Deviations in frequencies estimated using reduced-order models (MMDA and SNM) (Bhartiya and Sinha, 2011).

SNM analyses. The observation of deviations in frequencies in Figure 3.5.8 shows large values of frequency deviation for modes 73–89, whereas small deviations for modes 90–110 and then large and small frequency deviations intermixed for modes 111 to 120.

A closer look at the mode shapes of the mistuned bladed-disk assembly shows that modes 73–89 are blade-dominated torsional mode shapes with significant mode localization. Because the torsional mode shapes are sensitive to the changes in thicknesses of the blades, the mode shapes 73–89 of the mistuned bladed-disk assembly are significantly different from the mode shapes of the nominal tuned bladed-disk assembly, which results in large frequency deviations. On the other hand, for modes 90–110, lateral bending, torsion, and elongation modes are all present. A closer look at these mode shapes shows that the torsional mode shapes present in the range are disk dominated with small or no mode localization, hence they are not significantly altered by the mistuning. The elongation and lateral bending mode shapes present in the range are not disk dominated, but because the lateral bending and elongation mode shapes are not sensitive to the changes in the thicknesses of the blades, these mode shapes are also not altered due to mistuning. This results in small or no deviations in frequencies for modes 90–110. For modes 111 to 120, blade-dominated second bending modes (7th family) are also present along with disk-dominated torsional and elongation modes, all of which do not show significant mode localization, hence are similar to the nominal tuned bladed-disk assembly. Therefore, for torsional and elongation modes in the range, significant deviation in frequency is not observed. For modes corresponding to second bending in the range (modes 111, 115, and 117), although the mode shapes are not localized, they are blade dominated and because the natural frequency of the bending mode is sensitive to the thickness of the blade, significant shift in natural frequency is observed for modes corresponding to second bending modes. This phenomenon is similar to what is observed for the first bending family in Figures 3.5.5 and 3.5.6, where modes 1–12 are not localized but significant deviation in natural frequencies is observed. This analysis is also confirmed by the MAC values plotted for modes 73–120 for MMDA and frequency mistuning based on lateral bending, torsion, and elongation blade modes (Figure 3.5.9). As discussed earlier, frequency mistuning based on lateral bending or elongation modes does not capture geometric mistuning and the mode shapes estimated using an SNM analysis match with those of the nominal tuned system. Hence MAC values closer to 1 in Figures 3.5.9b and 3.5.9d suggest that the mistuned mode shapes are similar to the mode shapes of the nominal system, whereas MAC values closer to zero suggest that the mode shapes are significantly altered from the mode shapes of the nominal system. MAC values in Figure 3.5.9a show that MMDA is able to estimate the mistuned modes accurately.

Another mistuning pattern is created with two POD features. The first POD feature for the change in the thickness is same as that for the mistuning pattern number 1, that is,

$$\mathbf{u}_1 = [1 \quad 1 \quad 1 \quad 1 \quad 1 \quad 1]^T \tag{3.5.2}$$

whereas the second POD feature, Figure 3.5.10, represents a linear variation in the thickness, that is,

$$\mathbf{u}_2 = [-1 \quad -0.6 \quad -0.2 \quad 0.2 \quad 0.6 \quad 1]^T \tag{3.5.3}$$

Hence, the multiplying factor vector for the thickness change in each blade is represented as

$$\mathbf{u}_0 + \xi_{1i}\mathbf{u}_1 + \xi_{2i}\mathbf{u}_2; \quad i = 1, 2, \ldots, 24 \tag{3.5.4}$$

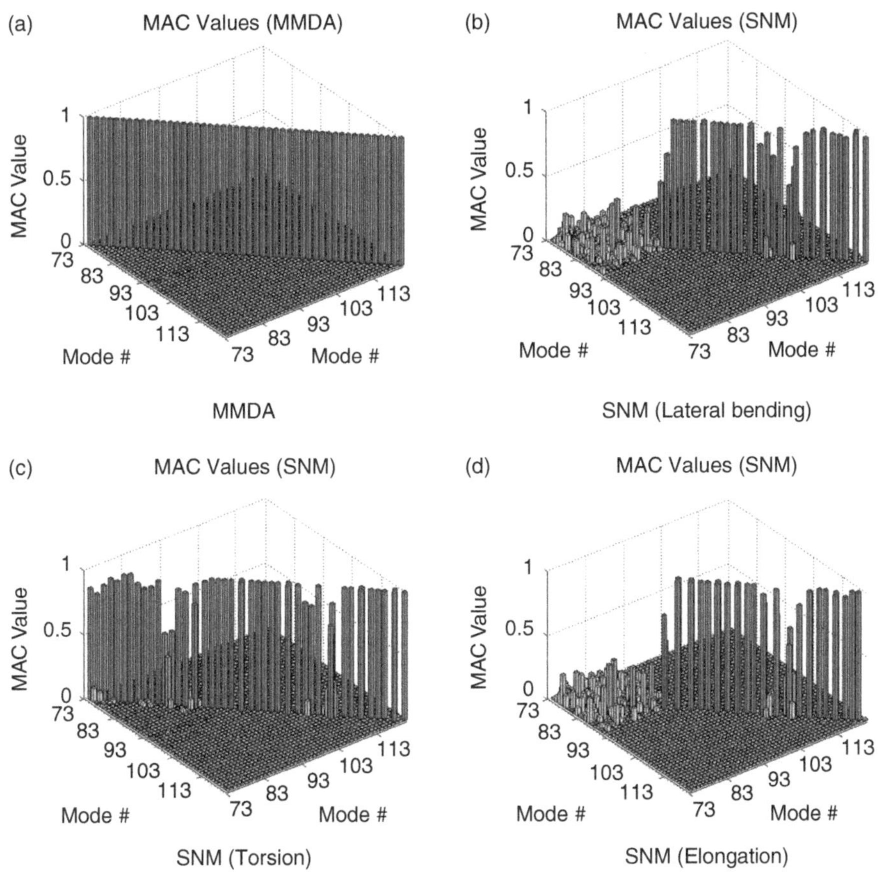



**Figure 3.5.9.**  MAC values for modes 73–120 calculated using MMDA, SNM (lateral bending), SNM (torsion), and SNM (elongation) (Bhartiya and Sinha, 2011).

where

$$\mathbf{u}_0 = \mathbf{u}_1 \tag{3.5.5}$$

The mistuning parameter values ($\xi_1$ and $\xi_2$) for POD # 1 and # 2 are given in Figures 3.5.3 and 3.5.11, respectively. The means of mistuning parameter values are almost zero and the standard deviations are 0.017 and 0.015, respectively. Note that mean ($\mu$) is not exactly zero because of a finite number of random variables. The maximum value of deviation in blade thickness is 3 percent of the average blade thickness.

A look at the values of mistuning parameters for each blade shows that for sectors 12 ($\xi_1 = 0.0061302$, $\xi_2 = -0.0186800$) and 15 ($\xi_1 = -0.029931$, $\xi_2 = 0.0045726$), mistuning parameters are dominated by POD 2 and 1, respectively, and the mistuning parameter values for the other POD feature (POD 1 for sector 12 and POD 2 for sector 15) are very small, that is, the mistuning values in sectors 12 and 15 are closely aligned to the directions of POD 2 and 1, respectively. Therefore the mode

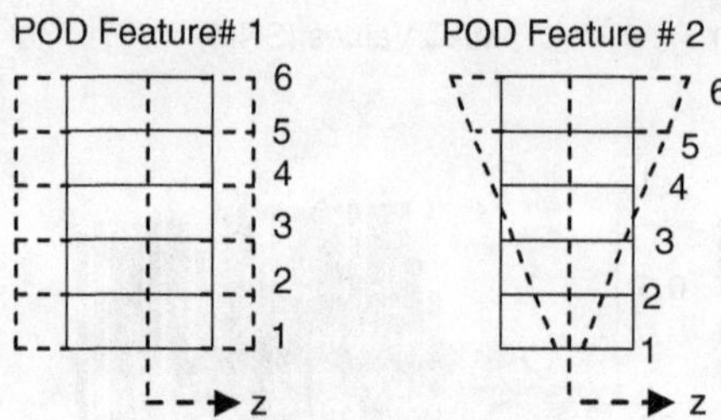


**Figure 3.5.10.** Blade thickness for each POD feature (Sinha, 2009).

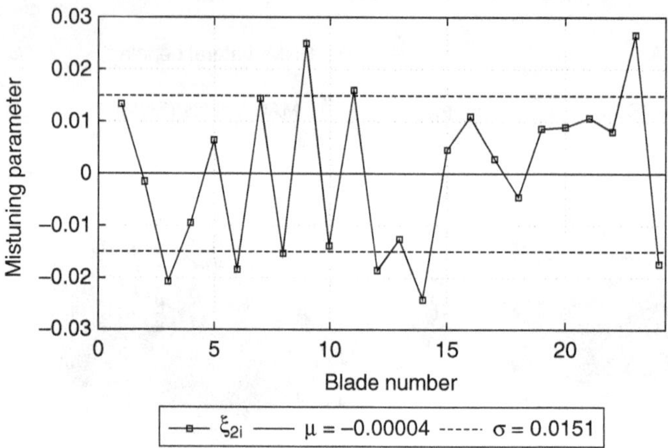


**Figure 3.5.11.** Mistuning pattern for POD # 2 (Bhartiya and Sinha, 2013b).

shapes ($\Phi_{a12}$ and $\Phi_{a15}$) from cyclic analyses of sectors 12 and 15 can be used to form the bases for the reduced-order model in Equation (3.4.35) as they will be close to mode shapes ($\Phi_2$ and $\Phi_1$) in Equation (3.4.10). Cyclic analyses are run for the two sectors 12 and 15 and mode shapes from the first five families are used in MMDA analysis.

Figure 3.5.12 plots deviations in frequencies estimated by MMDA and exact full rotor analysis (ANSYS) as given by Equations (3.5.6) and (3.5.7), respectively. Figure 3.5.13 plots the % errors in estimation as given by Equation (3.5.8), defined as follows:

$$Dev_{MMDA} = Freq_{MMDA} - Freq_{Tnd} \qquad (3.5.6)$$

$$Dev_{Act} = Freq_{Act} - Freq_{Tnd} \qquad (3.5.7)$$

$$Error D(\%) = \frac{Dev_{MMDA} - Dev_{Act}}{Dev_{Act}} * 100 \qquad (3.5.8)$$

where $Freq_{Tnd}$, $Freq_{MMDA}$, and $Freq_{Act}$ are natural frequencies of the tuned disk, the mistuned disk using MMDA, and the mistuned disk using full (360 degree) rotor

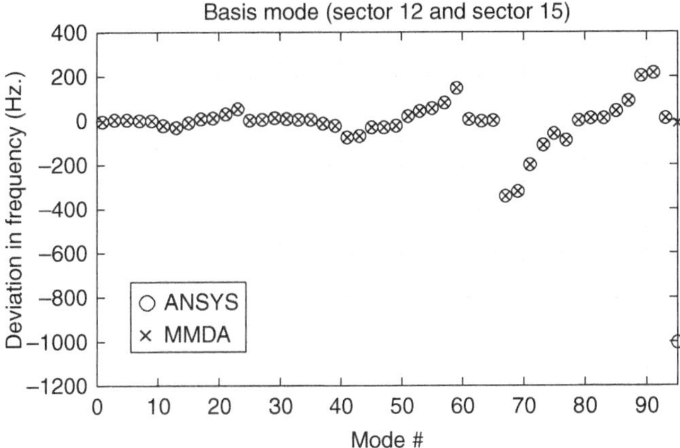



**Figure 3.5.12.** Deviations in natural frequencies estimated using full-rotor finite element model (FEM) and MMDA (alternative bases, sectors 12 and 15) (Bhartiya and Sinha, 2013b).

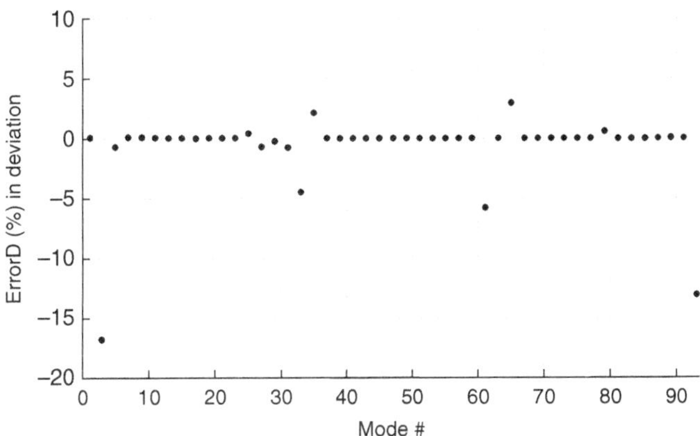



**Figure 3.5.13.** Errors (%) in deviations of natural frequencies estimated using MMDA (alternative bases, sectors 12 and 15) (Bhartiya and Sinha, 2013b).

analysis, respectively. As observed from these plots the estimates of natural frequencies from MMDA based on mode shapes from actual sectors as bases are very close to the actual values, which suggests that the use of alternative bases is valid for MMDA analysis. It should be noted that a few large values of $ErrorD(\%)$ are due to small values of corresponding $Dev_{Act}$ in Equation (3.5.8).

Next, MMDA with modified bases, Equation (3.4.36), is applied to bladed disk with extremely large mistuning. Mistuning is applied along the POD #1, that is, along the thicknesses of the blades. Mistuning parameter values along POD #1 are again given by Figure 3.5.3. Blade #23 is the rogue blade that has additional large mistuning along POD feature #2 (Figure 3.5.10) and #3, (Figure 3.5.14).

**Table 3.5.1.** Mistuning parameters for the rogue blade (blade #23) (Bhartiya and Sinha, 2013b)

| $\xi_1$ | $\xi_2$ | $\xi_3$ |
|---|---|---|
| −0.027854 | 0.0667100 | 0.0608370 |

POD Feature #3

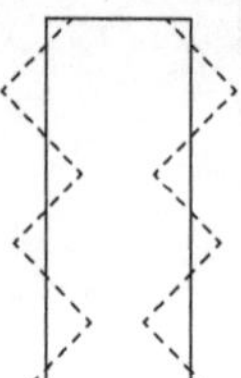

**Figure 3.5.14.** POD feature # 3 (Bhartiya and Sinha, 2013b).

POD feature 3 ($\mathbf{u}_3$) is created by taking the component of a specified vector $\mathbf{v}$ (Equation 3.5.9) so that it is orthogonal to both $\mathbf{u}_1$ and $\mathbf{u}_2$ using Gram–Schmidt ortho-normalization (Weisstein, 2006), that is,

$$\mathbf{v} = [1 \quad -1 \quad 1 \quad -1 \quad 1 \quad -1]^T \tag{3.5.9}$$

$$\mathbf{w} = \mathbf{v} - proj_{\mathbf{u}1}(\mathbf{v}) - proj_{\mathbf{u}2}(\mathbf{v}) \tag{3.5.10}$$

$$\mathbf{u}_3 = \frac{\mathbf{w}}{\|\mathbf{w}\|} \tag{3.5.11}$$

The values of mistuning parameters for the rogue blade are given in Table 3.5.1.

The mean value (excluding the rogue blade) of norms of perturbation vectors $\delta\mathbf{w}(\mathbf{p})$, Equation (3.4.4), is 0.0025, whereas the norm of perturbation vector for the rogue blade is 0.0145, that is, the perturbation in the rogue blade geometry is 5.8 times the average perturbation value.

MMDA analyses are run for the first family of the modes. To consider the impact of extremely large mistuning, first analysis is run without including the mode shapes of the sector with rogue blade, that is, only the "average" mode shapes and the mode shapes from geometry perturbed along POD feature 1 are included in the bases. Then the mode shapes from the rogue blade, $\Phi_{rogue}$, are also included in the MMDA analysis. Full-rotor ANSYS analysis is also run to compare the estimated natural frequencies with the true values. The deviations in frequencies and % errors in deviations, as given by Equations (3.5.6)–(3.5.8), are presented in Figures 3.5.15 and 3.5.16, respectively.

As discussed earlier, in the presence of extremely large mistuning, the inclusion of just the "average" modes and the modes from POD analysis is not sufficient to form suitable bases, as observed from the results of MMDA analysis with only the

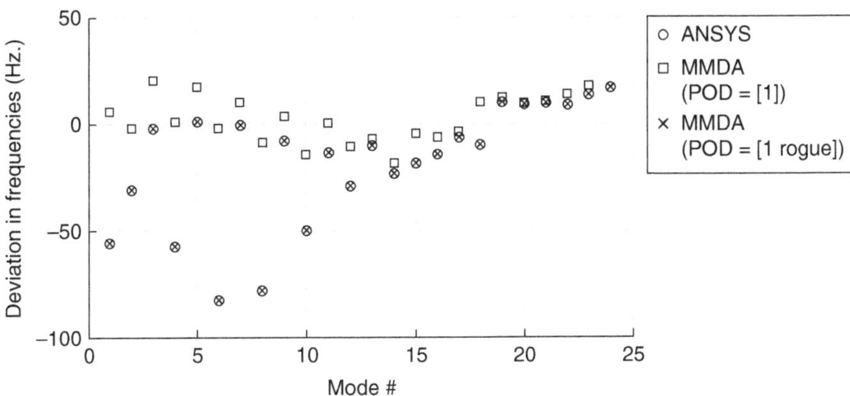

**Figure 3.5.15.** Deviations in natural frequencies estimated using full-rotor ANSYS, MMDA (POD 1 only), and MMDA (POD 1 and rogue blade) (Bhartiya and Sinha, 2013b).

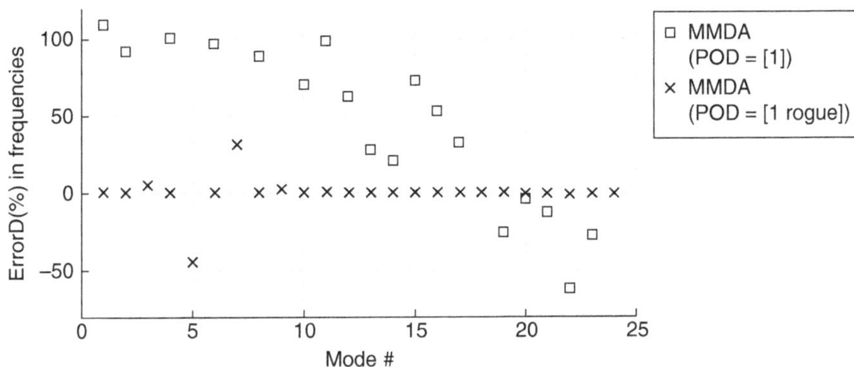

**Figure 3.5.16.** Errors (%) in deviations of natural frequencies estimated using MMDA (POD 1 only) and MMDA (POD 1 and rogue blade) (Bhartiya and Sinha, 2013b).

"average" and POD 1 modes in Figures 3.5.15 and 3.5.16. We can see large errors in the natural frequency estimates from the MMDA analysis without the rogue blade mode shapes. It can also be observed from Figures 3.5.15 and 3.5.16 that inclusion of mode shapes from the rogue blade rectifies this problem, and the natural frequency estimates from the MMDA analysis with rogue blade are very close to the true values.

## 3.5.2    Integrally Bladed Rotor

MMDA is demonstrated on the first stage rotor of a research Transonic Compressor (Vishwakarma et al., 2015) which is an eighteen airfoil integrally bladed rotor. Geometry of each blade is measured by CMM.

The finite element representation of the full (360°) rotor consists of 166,068 nodes and 181,800 elements of which are all linear hexahedral elements except for a small hex-tet transition layer at the disk rim to transition the mesh from the disk to airfoil geometries. The hexahedral elements use reduced integration with hourglass control so the analysis will not suffer from volumetric locking. Three hex elements are used through the thickness of the airfoil to further improve the models ability to accurately capture bending response. Typical Ti 6–4 material properties are used with 17.5M psi Young's Modulus, 0168 $lb/in^3$ density, and 0.3 Poisson's ratio. The model is constrained in all directions at the aft side of the flange located at the disk bore.

Special care was used to maintain consistent element topology for each sector so that numerical variations were not introduced from the sector discretization approach. A single sector of the disk and airfoil fillet was created, meshed, and then copied and rotated to build the 360-degree disk. The geometries of the as-measured airfoils (CMM data) were provided as consistent sets of coordinates defined as cross-sections and used to generate the surface and volume definition of each blade. With mesh seeding to define element numbers, hex meshing, and the copied disk sector, each sector was topologically consistent.

POD features from the bladed rotor are extracted by using the method developed by Sinha et al. (2008) and described here. Let $s$ be the number of nodes in the ANSYS model for a sector blade and spatial coordinates of $i^{th}$ node of blade # $\ell$ be denoted by $[px_{i\ell}\ py_{i\ell}\ pz_{i\ell}]$. Then, coordinates of all the $s$ nodes are arranged as a one-dimensional column vector, $\mathbf{w}$, of length $3s$, that is,

$$\mathbf{w}_\ell = [px_{1\ell} \quad \cdots \quad px_{s\ell} \quad py_{1\ell} \quad \cdots \quad py_{s\ell} \quad pz_{1\ell} \quad \cdots \quad pz_{s\ell}]^T \qquad (3.5.12)$$

The average geometry coordinates vector is obtained as

$$\bar{\mathbf{w}} = \frac{1}{18} \sum_{\ell=1}^{18} \mathbf{w}_\ell \qquad (3.5.13)$$

For each blade, deviations of nodal coordinates from their mean values are obtained:

$$\delta\mathbf{w}_\ell = \mathbf{w}_\ell - \bar{\mathbf{w}} \qquad (3.5.14)$$

Arranging these nodal coordinate deviations vectors as columns, the following matrix is formed:

$$\delta W = [\delta\mathbf{w}_1 \quad \delta\mathbf{w}_2 \quad \cdots \quad \delta\mathbf{w}_{18}] \qquad (3.5.15)$$

POD vectors for the mistuned rotor is obtained from SVD (singular value decomposition) (Sinha et al., 2008) of $\delta W$:

$$\delta W = U \Sigma V^T \qquad (3.5.16)$$

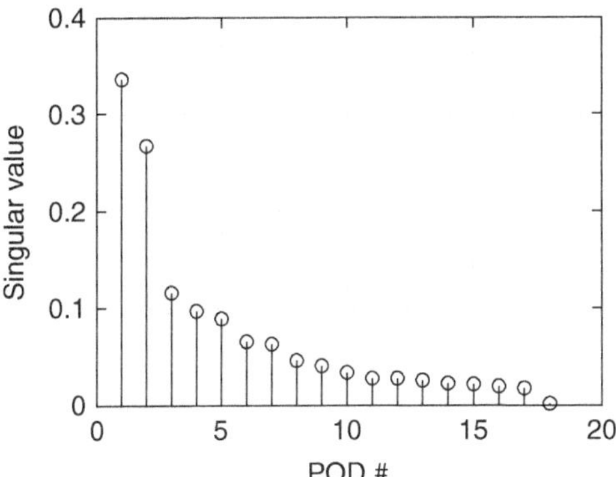

**Figure 3.5.17.** Singular values for integrally bladed rotor geometric mistuning (Vishwakarma et al., 2015).

where

$$U = [\mathbf{u}_1 \quad \mathbf{u}_2 \quad \cdots \quad \mathbf{u}_n] \tag{3.5.17}$$

$$\Sigma = diag[\sigma_1 \quad \sigma_2 \quad \cdots \quad \sigma_{18}] \tag{3.5.18}$$

$$V = [\mathbf{v}_1 \quad \mathbf{v}_2 \quad \cdots \quad \mathbf{v}_{18}] \tag{3.5.19}$$

The dimensions of matrices, $U$ and $V$, are $3s \times 18$ and $18 \times 18$, respectively. These matrices are composed of orthonormal sets of vectors. Vectors $\mathbf{u}_i$ are described as POD features, and singular values $\sigma_i$ are weights of these POD features. Equation (3.5.16) can also be written as

$$\delta W = \sum_{i=1}^{18} \mathbf{u}_i \sigma_i \mathbf{v}_i^T \tag{3.5.20}$$

More precisely, the contribution of ith POD feature to deviations in nodal coordinates of jth blade is given by $\mathbf{u}_i \sigma_i v_{ji}$ where $v_{ji}$ is the element of vector $\mathbf{v}_i$. Singular values are shown in Figure 3.5.17.

In Figure 3.5.18, a few POD features are shown in which scales on the black-white map represent the magnitude of spatial deviation of each node of the finite element mesh.

Figure 3.5.19 represents the nodal diameter map of the tuned rotor with each blade having the average geometry described by Equation (3.5.13). The first family of modes represents first bending modes, the second family of modes contains predominantly first torsion modes, the third family of modes predominantly second bending modes, that is, blades have one antinode in the middle of the blade, and the fourth family of modes is found to be a mix of second bending modes and torsion modes.

## Natural Frequencies and Mode Shapes
### MMDA Results

First task in implementing MMDA is to construct bases vectors described by Equation (3.4.10). The matrix $\Phi_0$ is obtained by the sector analysis with average

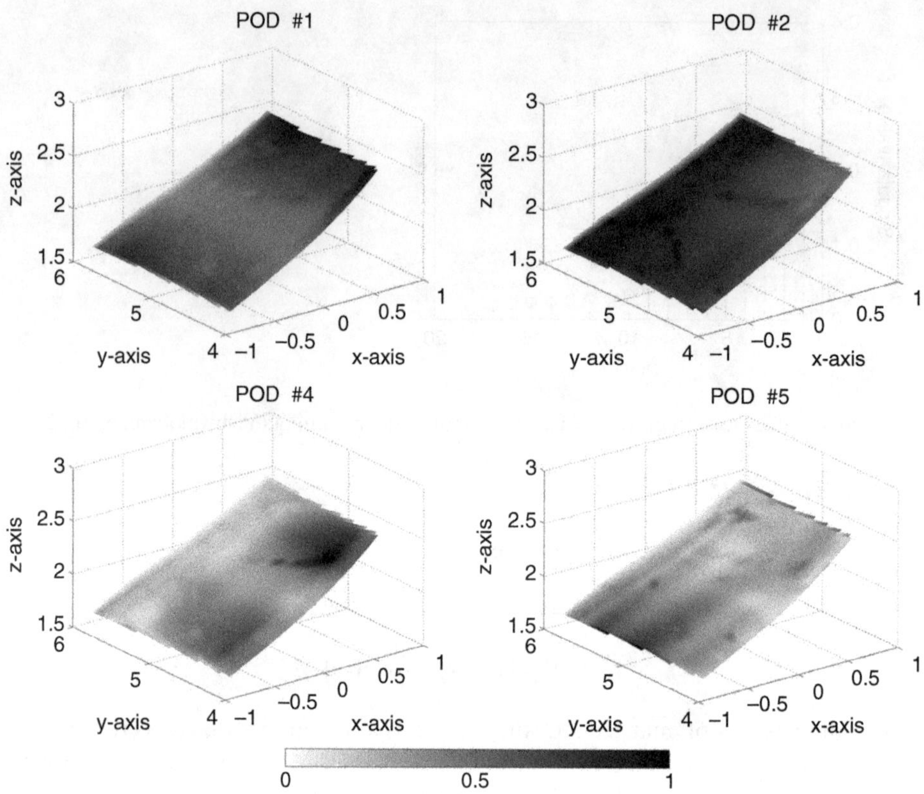

**Figure 3.5.18.** POD features for geometric mistuning of integrally bladed rotor (Vishwakarma et al., 2015).

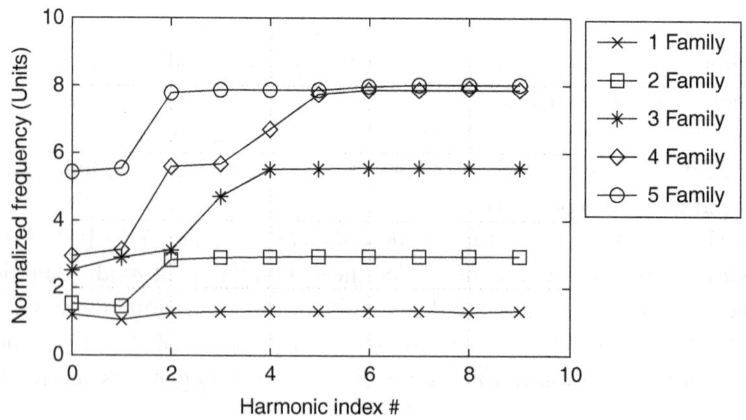

**Figure 3.5.19.** Nodal diameter map of rotor with each blade having "average" geometry (Vishwakarma et al., 2015).

blade geometry $\bar{\mathbf{w}}$. Matrices $\Phi_\ell$ are obtained by running sector analyses with the nodal coordinates of blades given by

$$\mathbf{w} = \bar{\mathbf{w}} + 0.5\sigma_\ell\mathbf{u}_\ell; \quad \ell = 1,2,\ldots,17 \tag{3.5.21}$$

It is recalled that matrices $\delta M$ and $\delta K$ are block diagonal:

$$\delta M = \begin{bmatrix} \delta M_1 & & & \\ & \delta M_2 & & \\ & & \ddots & \\ & & & \delta M_{18} \end{bmatrix} \tag{3.5.22}$$

and

$$\delta K = \begin{bmatrix} \delta K_1 & & & \\ & \delta K_2 & & \\ & & \ddots & \\ & & & \delta K_{18} \end{bmatrix} \tag{3.5.23}$$

where $\delta M_\ell$ and $\delta K_\ell$ are deviations in mass and stiffness matrices of sector # $\ell$ due to geometric mistuning. Mass and stiffness matrices are obtained for a sector with each blade for 0-degree interblade phase angle. Subtracting mass and stiffness matrices for sector with "average" blade for 0-degree interblade phase angle from these matrices, $\delta M_\ell$ and $\delta K_\ell$ are obtained without any approximations.

Natural frequencies and mode shapes are obtained from the reduced-order model for different numbers of POD features, and compared to those from full (360 degree) analysis of bladed rotor using ANSYS (see Figures 3.5.20–22).

MAC[17] plots (see Figure 3.5.21) indicate that only after the inclusion of the first nine POD features, does the accuracy of MMDA results become considerable. Deviations in natural frequencies are shown in Figure 3.5.20. With seventeen PODs, MMDA results are highly accurate (see Figure 3.5.22).

### *SNM and FMM Results*

Natural frequencies and mode shapes are also obtained for the first and second families from SNM (with first 90 modes) and FMM (with appropriate 18 modes). They are compared to full rotor (360 degrees) ANSYS results in the same way as done for MMDA (see Figures 3.5.23–3.5.25). Mistuned natural frequencies and mode shapes deviate from true values by large amounts for FMM. Even though SNM results are better, there are significant errors in mode shapes and natural frequencies. The errors in first-family FMM and SNM are notable as the first family is an isolated family of modes that was assumed for the development of both SNM and FMM (see Section 3.2). Surprisingly, mistuned frequencies and mode shapes for nonisolated second-family SNM are more accurate. The impact of these errors on the forced response will be presented next.

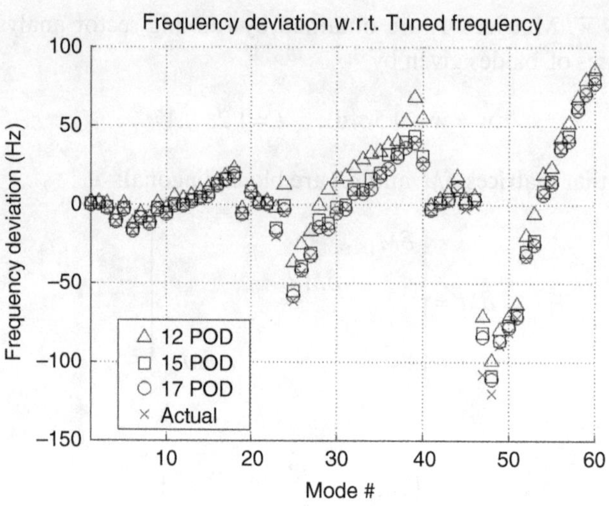

**Figure 3.5.20.** Frequency deviations from MMDA (12, 15, and 17 PODs) (Vishwakarma et al., 2015).

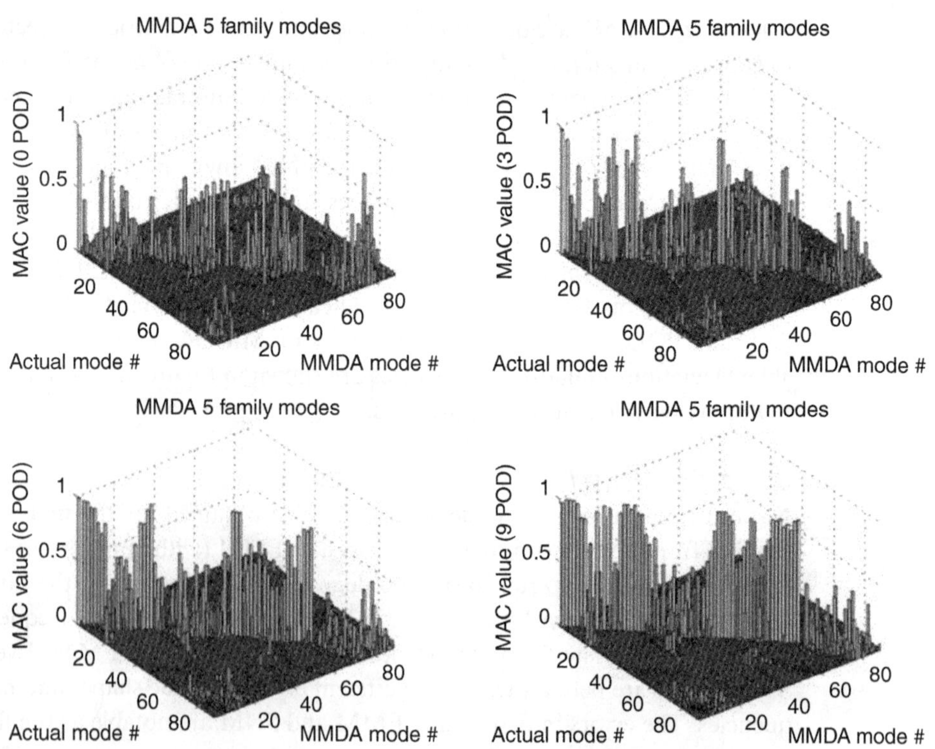

**Figure 3.5.21.** MAC for modes predicted by MMDA (0, 3, 6, and 9 PODs) (Vishwakarma et al., 2015).

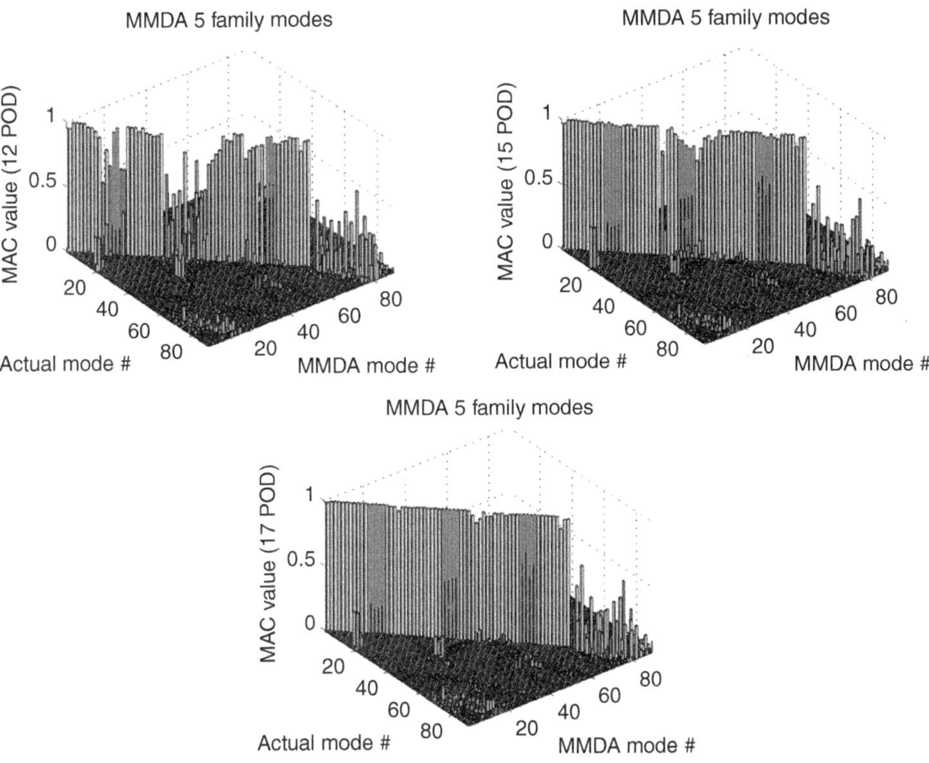



**Figure 3.5.22.** MAC for modes predicted by MMDA (12, 15, and 17 PODs) (Vishwakarma et al., 2015).

## Forced Response with Proportional Damping

Forced harmonic response of the bladed rotor based on the MMDA mode shapes is compared with that obtained using full rotor ANSYS mode shapes. Equation (3.4.14) is solved by first solving the reduced-order eigenvalue/eigenvector problem (Equation 3.4.18), and then using the mode superposition technique to get the harmonic response. From Equation (3.4.18),

$$K_r \Psi = M_r \Psi \Lambda_r \tag{3.5.24}$$

where

$$\Lambda_r = diag[\omega_1^2 \quad \omega_2^2 \quad \omega_3^2 \quad \cdots \quad \omega_{nr}^2] \tag{3.5.25}$$

$$\Psi = [\psi_1 \quad \psi_2 \quad \cdot \quad \cdot \quad \cdot \quad \psi_{nr}] \tag{3.5.26}$$

and *nr* is the number of modes from reduced order model. Let

$$\mathbf{y}(t) = \Psi \mathbf{z}(t) \tag{3.5.27}$$

Substituting Equation (3.5.27) into Equation (3.4.14) and premultiplying with $\Psi^H$,

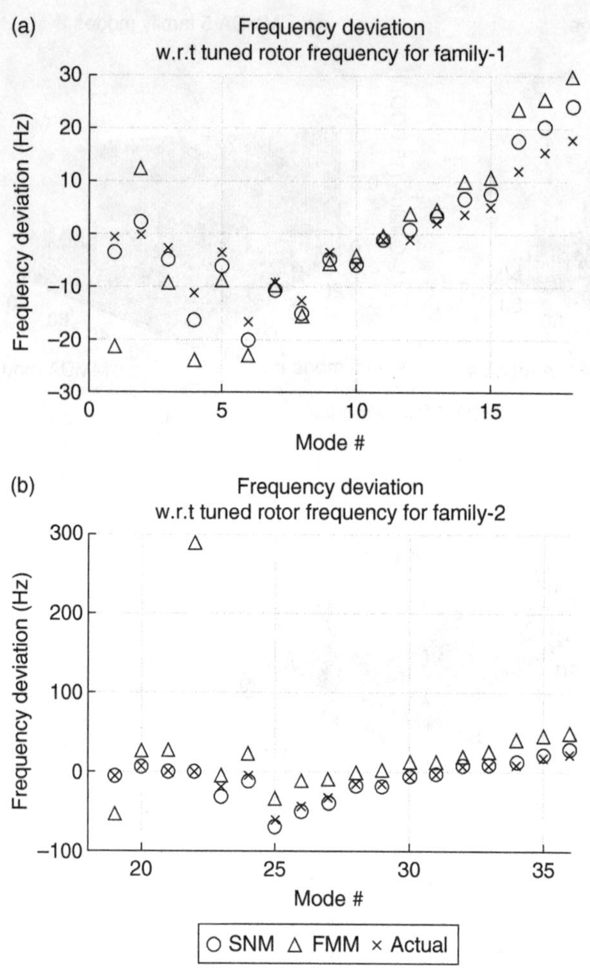


**Figure 3.5.23.** Frequency deviation comparison (SNM and FMM) (Vishwakarma et al., 2015).

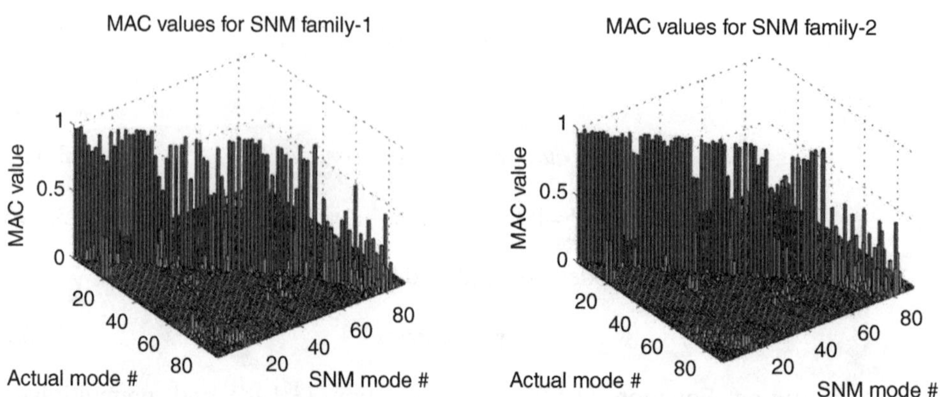


**Figure 3.5.24.** MAC values for SNM (first and second families of modes) (Vishwakarma et al., 2015).

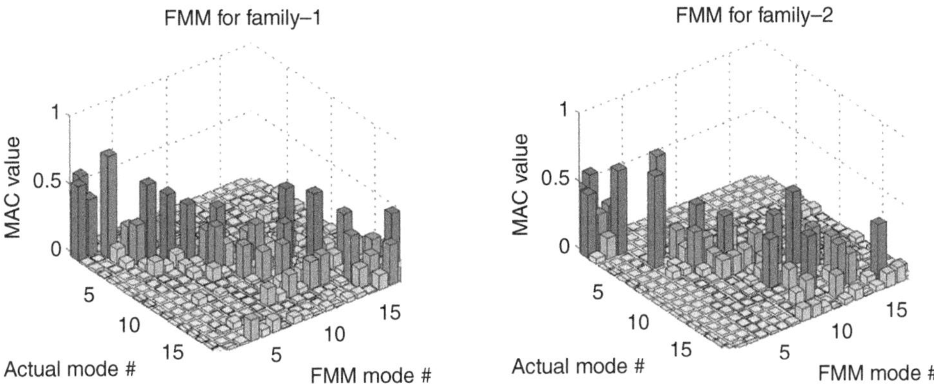


**Figure 3.5.25.** MAC values for FMM (first and second families of modes) (Vishwakarma et al., 2015).

$$\Psi^H M_r \Psi \ddot{\mathbf{z}} + \Psi^H \Phi^H C \Phi \Psi \dot{\mathbf{z}} + \Psi^H K_r \Psi \mathbf{z} = \Psi^H \Phi^H \mathbf{f}(t) \tag{3.5.28}$$

It is assumed that the damping matrix in Equation (3.5.28) has the Rayleigh form:

$$C = \alpha M + \beta K \tag{3.5.29}$$

where $\alpha$ and $\beta$ are proportionality constants. Using Equations (3.5.24)–(3.5.26), the following set of decoupled modal equations is obtained:

$$\ddot{\mathbf{z}} + (\alpha I_{nr} + \beta \Lambda_r) \dot{\mathbf{z}} + \Lambda_r \mathbf{z} = \Psi^H \Phi^H \mathbf{f}(t) \tag{3.5.30}$$

For the pth engine-order excitation,

$$\mathbf{f}(t) = \mathbf{f}_0 e^{j\omega t} \tag{3.5.31}$$

$$\mathbf{f}_0 = \begin{bmatrix} \mathbf{f}_1 & \mathbf{f}_1 e^{-j\varphi} & \mathbf{f}_1 e^{-j2\varphi} & \cdots & \mathbf{f}_1 e^{-j(n-1)\varphi} \end{bmatrix}^H \tag{3.5.32}$$

where

$$\varphi = \frac{2\pi p}{n} \tag{3.5.33}$$

and $\mathbf{f}_1$ is the force magnitude row vector for each sector. The steady-state response is given by

$$\mathbf{z}(t) = \mathbf{z}_a e^{j\omega t} \tag{3.5.34}$$

Substituting Equation (3.5.34) into Equation (3.5.30) and equating the coefficients of $e^{j\omega t}$ on both sides,

$$[-\omega^2 I_{nr} + (\alpha I_{nr} + \beta \Lambda_r) j\omega + \Lambda_r] \mathbf{z}_a = \mathbf{g} \tag{3.5.35}$$

where

$$\mathbf{g} = \Psi^H \Phi^H \mathbf{f}_0 \tag{3.5.36}$$

Because of the decoupled nature of the system of Equations (3.5.35), it is straightforward to obtain

$$z_{ai} = \frac{g_i}{(-\omega^2 + \omega_i^2) + j\omega(\alpha + \beta\omega_i^2)} \tag{3.5.37}$$

where $z_{ai}$ and $g_i$ are ith elements of vectors $\mathbf{z}_a$ and $\mathbf{g}$, respectively. Lastly, the complex steady-state amplitude of the physical displacement vector, $\mathbf{x}_a$, is obtained using transformations (3.4.9) and (3.5.27) as

$$\mathbf{x}_a = \Phi\Psi\mathbf{z}_a \tag{3.5.38}$$

Further, it is assumed that $\alpha = 0$ in Equation (3.5.29) and $\beta$ is chosen such that the damping ratio for the resonant tuned mode equals 0.001. It is again observed that the forced harmonic response of the bladed rotor based on the MMDA converges to that based on ANSYS as the number of POD features increase in the construction of matrix $\Phi$ (see Equation (3.4.10)). The results for MMDA and SNM are compared with those obtained on the basis of ANSYS mistuned modes, for example, Figures 3.5.26 and 3.5.27 corresponding to second engine order excitation and modal families 1 and 2, respectively. Many excitation frequencies are chosen within ±3 percent of the first- and second-family natural frequencies corresponding to the second nodal diameter of the tuned bladed disk. Note that normalized maximum amplitudes (*nma*) in Figures 3.5.26 and 3.5.27 are computed at each frequency as follows:

$$nma = \frac{\|\mathbf{x}_a\|_\infty}{\max_\omega \|\mathbf{x}_{at}\|_\infty} \tag{3.5.39}$$

where $\mathbf{x}_a$ and $\mathbf{x}_{at}$ are steady-state amplitude vectors for mistuned and tuned rotors, respectively.

Because the time required to do the MMDA analysis is much smaller as compared to full-rotor ANSYS, it can also be used to perform quick Monte Carlo analysis to capture the statistics of forced response.

## Statistics of Forced Response

To generate the statistical distribution of peak maximum amplitudes, the Monte Carlo method in which coefficients of POD features will be randomly chosen can be used, and then MMDA can be used to find the peak maximum amplitude. Repeating this process a large number of times, the probability distribution function can be generated. However, for each simulation, deviations in mass and stiffness matrices must be computed by $n$ (number of blades) FE sector analyses, which

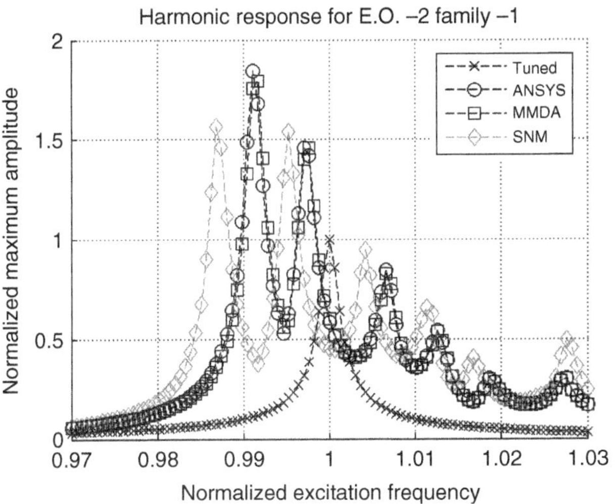


**Figure 3.5.26.** Forced harmonic response comparison (family 1) (Vishwakarma et al., 2015).

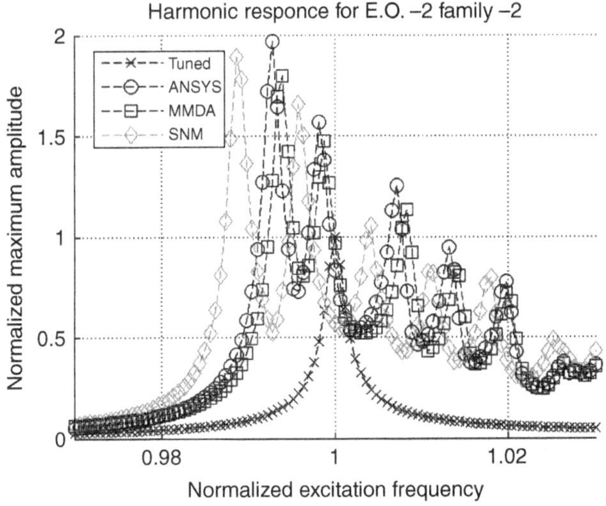


**Figure 3.5.27.** Forced harmonic response comparison (family 2) (Vishwakarma et al., 2015).

is time consuming. In a recent paper, it has been proposed that deviations in mass and stiffness matrices can be computed using a second-order Taylor series expansion for which all the partial derivatives or sensitivities are precomputed using a finite difference scheme. The accuracy of this approach has been shown (Bhartiya and Sinha, 2013a) for an academic rotor with two POD features. But, this approach is under further development for a real IBR with real CMM data having a large number of important POD features particularly in view of the possibility of inaccuracies in partial derivatives computation using the finite difference.

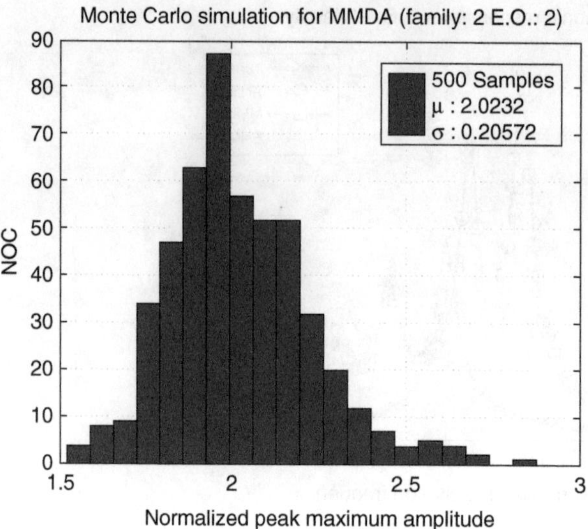


**Figure 3.5.28.** Monte Carlo simulation with 500 permutations of blades (MMDA) (Vishwakarma et al., 2015).

An efficient process to generate the statistics of peak maximum amplitudes through random permutations of blades has been developed (Vishwakarma and Sinha, 2015). A new IBR can be simulated by randomly rearranging the blades on the rotor. Then the peak maximum amplitude for this rotor can be calculated using MMDA in a computationally efficient manner without requiring any new FE sector analysis. Repeating this process for a large number of random permutations, the probability distribution function of peak maximum amplitudes can be generated. The number of possible permutations of $n$ blades in a rotor is $(n-1)!/2$, which can be a large number. For example, when $n = 18$, $(n-1)!/2 = 1.7784 \times 10^{14}$. All these random permutations are also samples for aforementioned Monte Carlo simulations. Vishwakarma and Sinha (2015) have shown that probability distribution functions of peak maximum amplitudes from random permutation simulations and the conventional Monte Carlo simulations are almost the same.

To capture the statistics of a full-rotor forced peak harmonic response, a Monte Carlo simulation with MMDA is done. MMDA requires a one- time preprocessing time for reduced-order matrix generation for a given set of blades, and those blades then could be rearranged in various permutations to perform Monte Carlo simulation for forced harmonic response without any new finite element analysis (Sinha and Bhartiya, 2010). Monte Carlo simulation has been performed for the bladed rotor with 500 samples of random permutations of blades, second engine order excitation of the second family of modes. Figure 3.5.28 presents the distribution of normalized peak maximum amplitude (*npma*), which is computed over all blades and a range (±3 percent of mean value) of excitation frequencies. Specifically, *npma* is computed as follows:

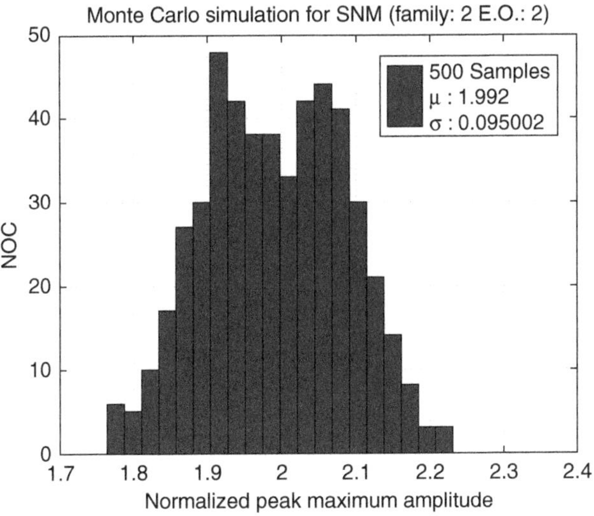


**Figure 3.5.29.** Monte Carlo simulation with 500 permutations of blades (SNM) (Vishwakarma et al., 2015).

$$npma = \frac{\max\limits_{\omega} \left\| \mathbf{x}_a \right\|_\infty}{\max\limits_{\omega} \left\| \mathbf{x}_{at} \right\|_\infty} \tag{3.5.40}$$

where $\mathbf{x}_a$ and $\mathbf{x}_{at}$ are steady-state amplitude vectors for mistuned and tuned rotors, respectively. Also, NOC stands for the number of occurrences. The distribution of *npma* is also obtained for SNM with same permutations of blades as used in MMDA and is presented in Figure 3.5.29. It is clear that the distribution of *npma* from MMDA is quite different from those from SNM distribution. The standard deviation of MMDA distribution is almost two times that of SNM. Because the MMDA forced harmonic response is almost as accurate as full-rotor ANSYS, it is concluded that the Monte Carlo simulation results

from SNM are not correct. In Figure 3.5.30, both SNM and MMDA values of *npma* are presented for each permutation of blade arrangement.

## 3.6  Identification of Geometric Mistuning from Measured Data

Feiner and Griffin (2004a, 2004b) have developed frequency mistuning identification techniques for entire bladed disk assemblies by making many ad hoc approximations. This algorithm has been presented in Section 3.2.2. Mignolet, Delor, Rivas-Guerra (1999a, 1999b) and Lim, Pierre, and Castanier (2004) have also developed mistuning identification algorithms. But, these identification techniques are also based on reduced-order models representing frequency mistuning only.

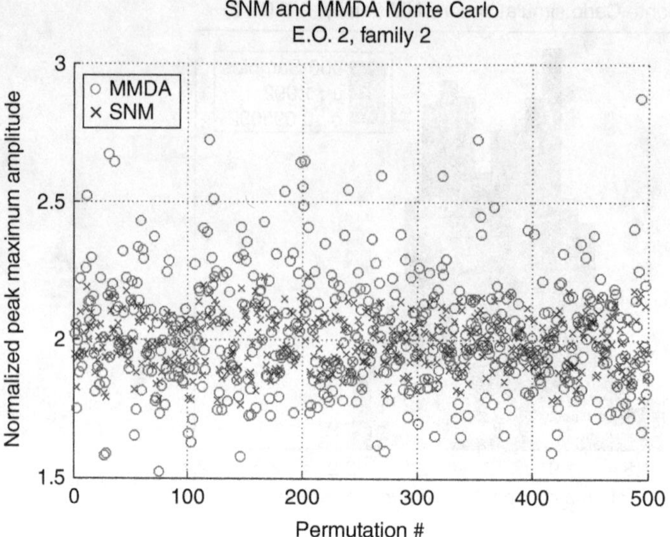



**Figure 3.5.30.** Normalized peak maximum amplitude for each permutation of blades (Vishwakarma et al., 2015).

Because the actual mistuning in an integrally bladed rotor occurs due to perturbations in blades' geometries, the identification technique based on frequency mistuning can lead to erroneous estimates of the bladed-disk modal characteristics, which in turn provoke large discrepancies in the prediction of the forced response of the entire system near or at resonance in the presence of a small damping ratio. It has been clearly shown that the frequency mistuning approach is not an accurate representation of geometric mistuning (Bhartiya and Sinha, 2011). Therefore, Bhartiya and Sinha (2014) have developed algorithms to identify coefficients of POD features representing geometric mistuning on the basis of modal and forced harmonic response data as well. Here, algorithms based on the knowledge of modal data will be presented.

As shown by Bhartiya and Sinha (2013a), the perturbations in mass and stiffness matrices of each sector can be represented using second-order Taylor series expansion. For example, for deviation in a sector's mass matrix,

$$\delta M_l = \sum_{s=1}^{np} \frac{\partial \delta M}{\partial \xi_s} \xi_{s,l} + \sum_{s=1}^{np} \sum_{t=1}^{np} \frac{\partial^2 \delta M}{\partial \xi_s \partial \xi_t} \frac{\xi_{t,l} \xi_{s,l}}{2} \tag{3.6.1}$$

where $\xi_{s,l}$ is the mistuning parameter for POD # $s$ and sector # $l$. Using this approximation, the $(v+1, \rho+1)$ element of $\Phi_i^H \delta M \Phi_j$ can be written as:

$$\sum_{s=1}^{np} \varphi_{i,1,v}^H \frac{\partial \delta M}{\partial \xi_s} \varphi_{j,1,\rho} \bar{\xi}(k) + \cdots \frac{1}{n} \sum_{s=1}^{np} \sum_{t=1}^{np} \varphi_{i,1,v}^H \frac{1}{2} \frac{\partial^2 \delta M}{\partial \xi_s \partial \xi_t} \varphi_{j,1,\rho} Z_t^T P(k) Z_s \tag{3.6.2}$$

$\overline{\xi}_s(k)$ is the $k^{th}$ discrete Fourier transform (Oppenheim and Shaeffer, 1975) of $\xi_{s,l}$ defined by:

$$\overline{\xi}_s(k) = \sum_{l=0}^{n-1} e^{-\frac{2\pi \iota}{n} l(v-\rho)} \xi_{s,l} \quad \text{and} \quad k = v - \rho \tag{3.6.3}$$

where $n$ is the number of blades in the disk. The expressions for $Z_s$ and $P(k)$ are provided by:

$$Z_t = \begin{bmatrix} \overline{\xi}_t(0) & \cdots & \overline{\xi}_t(n-1) \end{bmatrix}^T \tag{3.6.4}$$

$$P(k) = \begin{pmatrix} 0 & \cdots & 1 & \cdots & 0 \\ & \ddots & & & \\ 1 & 0 & 0 & 0 & \cdots & 0 \\ 0 & \cdots & 0 & 0 & \cdots & 1 \\ & & & \ddots & \\ 0 & \cdots & 0 & 1 & \cdots & 0 \end{pmatrix} \text{or}$$

$$P_{ij}(k) = \begin{matrix} \delta_{i-k,j} & i > k \\ \delta_{i-k+n,j} & i \le k \end{matrix} \tag{3.6.5}$$

where $\delta_{p,j} = 1$ when $p = j$, and 0 otherwise.

Similarly, the $(v+1, \rho+1)$ element of $\Phi_i^H \delta K \Phi_j$ can be written as:

$$\sum_{s=1}^{np} \varphi_{i,1,v}^H \frac{\partial \delta K}{\partial \xi_s} \varphi_{j,1,\rho} \overline{\xi}(k) + \quad \cdots \quad \frac{1}{n} \sum_{s=1}^{np} \sum_{t=1}^{np} \varphi_{i,1,v}^H \frac{1}{2} \frac{\partial^2 \delta K}{\partial \xi_s \partial \xi_t} \varphi_{j,1,\rho} Z_t^T P(k) Z_s \tag{3.6.6}$$

All the partial derivatives in Equations (3.6.1) and (3.6.6) are evaluated using a finite difference scheme (Bhartiya and Sinha, 2013a)

## Identification Using Measured Mistuned Frequencies and Mode Shapes

The eigenvalue problem for the reduced-order system for the $p^{th}$ mode can then be written as:

$$(\Phi^H K_t \Phi - \omega_p^2 \Phi^H M_t \Phi)\beta_p + (\Phi^H \delta K \Phi - \omega_p^2 \Phi^H \delta M \Phi)\beta_p = 0 \tag{3.6.7}$$

Because $\Phi$, $K_t$, and $M_t$ are defined by the mean geometry and a set of POD features of the system, they are known. The natural frequencies and mode shapes are measured; hence, $\omega_p$ and $\beta_p$ are also known. Therefore, the expression $(\Phi^H K_t \Phi - \omega_p^2 \Phi^H M_t \Phi)\beta_p$ can be directly calculated from the known values.

Let $\beta_p$ be the vector of modal coefficients of the known $p^{th}$ mode shape. Then $\beta_p$ can be represented as:

$$\beta_p = \begin{bmatrix} \beta_{p,0} \\ \vdots \\ \beta_{p,np} \end{bmatrix} \tag{3.6.8}$$

where $\beta_{p,j}$ is the modal coefficient vector corresponding to $\Phi_j$ in Equation (3.4.10). Further, $\beta_{p,j}$ consists of coefficients corresponding to mode shapes from sector analyses for different harmonic indices and can be represented as:

$$\beta_{p,j} = \begin{pmatrix} \beta_{p,j,0} \\ \beta_{p,j,1} \\ \vdots \\ \beta_{p,j,n-1} \end{pmatrix} \tag{3.6.9}$$

Here $\beta_{p,j,l}$ is the coefficient of $p^{th}$ mode corresponding to tuned mode shapes of average or geometry perturbed along a POD feature (represented by $j$) and harmonic index #$l$. The harmonic indices $l$ and $n - l$ correspond to traveling waves moving in opposite directions (say clockwise and counter clockwise, or vice versa) for repeated eigenvalues.

Using Equation (3.6.8), we can write:

$$\Phi^H \delta M \Phi \beta_p = \begin{pmatrix} \displaystyle\sum_{j=0}^{np} \Phi_0^H \delta M \Phi_j \beta_{p,j} \\ \vdots \\ \displaystyle\sum_{j=0}^{np} \Phi_{np}^H \delta M \Phi_j \beta_{p,j} \end{pmatrix} \tag{3.6.10}$$

Similar expression can be written for $\Phi^H \delta K \Phi \beta_p$. Using Equation (3.6.2), the expression for $\Phi_i^H \delta M \Phi_j \beta_{p,j}$ can be written as:

$$
\begin{aligned}
\Phi_i^H \delta M \Phi_j \beta_{p,j} = &\sum_{s=1}^{np}
\begin{pmatrix}
A(s)_{0,0}^{i,j}\,\overline{\xi}_s(0) & A(s)_{0,1}^{i,j}\,\overline{\xi}_s(n-1) & \cdots & A(s)_{0,n-1}^{i,j}\,\overline{\xi}_s(1) \\
A(s)_{1,0}^{i,j}\,\overline{\xi}_s(1) & A(s)_{1,1}^{i,j}\,\overline{\xi}_s(0) & \cdots & A(s)_{1,n-1}^{i,j}\,\overline{\xi}_s(2) \\
\vdots & \vdots & \cdots & \vdots \\
A(s)_{N-1,0}^{i,j}\,\overline{\xi}_s(n-1) & A(s)_{n-1,1}^{i,j}\,\overline{\xi}_s(n-2) & \cdots & A(s)_{n-1,n-1}^{i,j}\,\overline{\xi}_s(0)
\end{pmatrix}
\begin{pmatrix} \beta_{p,j,0} \\ \beta_{p,j,1} \\ \vdots \\ \beta_{p,j,n-1} \end{pmatrix} \\[2ex]
&+ \sum_{s=1}^{np}\sum_{t=s}^{np} \frac{1}{n}
\begin{pmatrix}
B(s,t)_{0,0}^{i,j} Z_t^T P(0) Z_s & B(s,t)_{0,1}^{i,j} Z_t^T P(n-1) Z_s & \cdots & B(s,t)_{0,n-1}^{i,j} Z_t^T P(1) Z_s \\
B(s,t)_{1,0}^{i,j} Z_t^T P(1) Z_s & B(s,t)_{1,1}^{i,j} Z_t^T P(0) Z_s & \cdots & B(s,t)_{1,n-1}^{i,j} Z_t^T P(2) Z_s \\
\vdots & \vdots & \cdots & \vdots \\
B(s,t)_{n-1,0}^{i,j} Z_t^T P(n-1) Z_s & B(s,t)_{n-1,1}^{i,j} Z_t^T P(n-2) Z_s & \cdots & B(s,t)_{n-1,n-1}^{i,j} Z_t^T P(0) Z_s
\end{pmatrix}
\begin{pmatrix} \beta_{p,j,0} \\ \beta_{p,j,1} \\ \vdots \\ \beta_{p,j,n-1} \end{pmatrix}
\end{aligned}
\tag{3.6.11}
$$

where

$$A(s)_{v,\rho}^{i,j} = \varphi_{i,1,v}^H \frac{\partial \delta M}{\partial \xi_s} \varphi_{j,1,\rho} \tag{3.6.12}$$

$$B(s,t)^{i,j}_{v,\rho} = \varphi^H_{i,1,v} \frac{1}{2} \frac{\partial^2 \delta M}{\partial \xi_s \partial \xi_t} \varphi_{j,1,\rho} \tag{3.6.13}$$

After a little algebra, Equation (3.6.11) can be written as:

$$\Phi^H_i \delta M \Phi_j \beta_{p,j} = \sum_{s=1}^{np} \hat{A}^p_{i,j}(s) Z_s + \sum_{s=1}^{np}\sum_{t=s}^{np} \hat{B}^p_{i,j}(s,t) \begin{pmatrix} Z^T_t P(0) \\ Z^T_t P(1) \\ \vdots \\ Z^T_t P(n-1) \end{pmatrix} Z_s \tag{3.6.14}$$

$$\hat{A}^p_{i,j}(s) = \begin{pmatrix} A(s)^{i,j}_{0,0}\beta_{p,j,0} & A(s)^{i,j}_{0,n-1}\beta_{p,j,n-1} & \cdots & A(s)^{i,j}_{0,1}\beta_{p,j,1} \\ A(s)^{i,j}_{1,1}\beta_{p,j,1} & A(s)^{i,j}_{1,0}\beta_{p,j,0} & \cdots & A(s)^{i,j}_{1,2}\beta_{p,j,2} \\ \vdots & \vdots & \cdots & \vdots \\ A(s)^{i,j}_{n-1,n-1}\beta_{p,j,n-1} & A(s)^{i,j}_{n-1,n-2}\beta_{p,j,n-2} & \cdots & A(s)^{i,j}_{n-1,0}\beta_{p,j,0} \end{pmatrix} \tag{3.6.15}$$

$$\text{and } \hat{B}^p_{i,j}(s,t) = \frac{1}{n} \begin{pmatrix} B(s,t)^{i,j}_{0,0}\beta_{p,j,0} & B(s,t)^{i,j}_{0,n-1}\beta_{p,j,n-1} & \cdots & B(s,t)^{i,j}_{0,1}\beta_{p,j,1} \\ B(s,t)^{i,j}_{1,1}\beta_{p,j,1} & B(s,t)^{i,j}_{1,0}\beta_{p,j,0} & \cdots & B(s,t)^{i,j}_{1,2}\beta_{p,j,2} \\ \vdots & \vdots & \cdots & \vdots \\ B(s,t)^{i,j}_{n-1,n-1}\beta_{p,j,n-1} & B(s,t)^{i,j}_{n-1,n-2}\beta_{p,j,n-2} & \cdots & B(s,t)^{i,j}_{n-1,0}\beta_{p,j,0} \end{pmatrix} \tag{3.6.16}$$

Therefore

$$\sum_{j=0}^{np} \Phi^H_i \delta M \Phi_j \beta_{p,j} = \sum_{s=1}^{np} \hat{A}^p_i(s) Z_s + \sum_{s=1}^{np}\sum_{t=1}^{np} \hat{B}^p_i(s,t) \begin{pmatrix} Z^T_t P(0) \\ Z^T_t P(1) \\ \vdots \\ Z^T_t P(n-1) \end{pmatrix} Z_s \tag{3.6.17}$$

where

$$\hat{A}^p_i(s) = \sum_{j=0}^{n-1} \hat{A}^p_{i,j}(s) \text{ and} \tag{3.6.18}$$

$$\hat{B}^p_i(s,t) = \sum_{j=0}^{np} \hat{B}^p_{i,j}(s,t) \tag{3.6.19}$$

Equation (3.6.17) then can be written in matrix form as:

$$\sum_{j=0}^{np} \Phi^H_i \delta M \Phi_j \beta_{p,j} = \Psi^M_{i,p} \begin{pmatrix} Z_1 \\ \vdots \\ Z_{np} \end{pmatrix} \tag{3.6.20}$$

where

$$\Psi_{i,p}^{M} = \left( \Psi_{i,p,1}^{M} \quad \cdots \quad \Psi_{i,p,np}^{M} \right) \text{ and} \tag{3.6.21}$$

$$\Psi_{i,p,s}^{M} = \hat{A}_{i}^{p}(s) + \sum_{t=1}^{np} \hat{B}_{i}^{p}(s,t) \begin{pmatrix} Z_{t}^{T} P(0) \\ Z_{t}^{T} P(1) \\ \vdots \\ Z_{t}^{T} P(n-1) \end{pmatrix} \tag{3.6.22}$$

An equation similar to Equation (3.6.20) can be generated for $\sum_{j=0}^{np} \Phi_{i}^{H} \delta K \Phi_{j} \beta_{p,j}$:

$$\sum_{j=0}^{np} \Phi_{i}^{H} \delta K \Phi_{j} \beta_{p,j} = \Psi_{i,p}^{K} \begin{pmatrix} Z_{1} \\ \vdots \\ Z_{np} \end{pmatrix} \tag{3.6.23}$$

From Equations (3.6.20) and (3.6.23),

$$\sum_{j=0}^{np} (\Phi_{i}^{H} \delta K \Phi_{j} - \omega_{p}^{2} \Phi_{i}^{H} \delta M \Phi_{j}) \beta_{p} = \left( \Psi_{i,p}^{K} - \omega_{p}^{2} \Psi_{i,p}^{M} \right) \begin{pmatrix} Z_{1} \\ \vdots \\ Z_{r} \end{pmatrix} \tag{3.6.24}$$

Hence the eigenvalue problem from Equation (3.6.7) can be written as:

$$\Psi_{p} \begin{pmatrix} Z_{1} \\ \vdots \\ Z_{np} \end{pmatrix} = \Gamma_{p} \tag{3.6.25}$$

where

$$(\omega_{p}^{2} \Phi^{H} M_{t} \Phi - \Phi^{H} K_{t} \Phi) \beta_{p} = \omega_{p}^{2} \Gamma_{p}^{M} - \Gamma_{p}^{K} = \Gamma_{p} \tag{3.6.26}$$

and

$$\Psi_{p} = \begin{pmatrix} \Psi_{0,p}^{K} \\ \vdots \\ \Psi_{np,p}^{K} \end{pmatrix} - \omega_{p}^{2} \begin{pmatrix} \Psi_{0,p}^{M} \\ \vdots \\ \Psi_{np,p}^{M} \end{pmatrix} \tag{3.6.27}$$

If $k$ modes are used then Equation (3.6.25) for each mode can be stacked and the complete set of equations can be written as:

$$\begin{pmatrix} \Psi_{p} \\ \vdots \\ \Psi_{p+k} \end{pmatrix} \begin{pmatrix} Z_{1} \\ \vdots \\ Z_{np} \end{pmatrix} = \begin{pmatrix} \Gamma_{p} \\ \vdots \\ \Gamma_{p+k} \end{pmatrix} \tag{3.6.28}$$

Equation (3.6.28) can be used to calculate the values of mistuning parameters. Because Equation (3.6.28) is nonlinear ($\psi_q$ is a function of $(Z_1^T \cdots Z_{np}^T)^T$), an iterative least squares solution can be used to solve the system of equations. The initial solution for the iterative procedure can be obtained by taking only the constant term in Equation (3.6.22), that is, $\Psi_{i,p,s}^M = \hat{A}_i^p(s)$.

## Numerical Examples

The bladed disk (see Figure 3.5.1) is considered again. The blades are mistuned by changing the thickness (POD # 1) and surface inclination (POD # 2) of the blades as shown in Figure 3.5.10. Specifications of POD feature # 1, $\xi_{1l}$, and #2, $\xi_{2l}$, uniquely determine blade thicknesses $b_l$, at locations $p = 1, 2, 3, 4, 5,$ and 6 for blade # $l$ in Figure 3.5.10 as described by Equation (3.5.4).

The actual mistuned bladed disk is created by picking the values of mistuning parameters for the blade thickness and surface inclination from normal distributions with zero mean and standard deviations of 1.7 percent and 1.5 percent, respectively. Modal analysis of this disk is performed for the first forty-eight modes using full-rotor (360 degrees) analysis in ANSYS. The mode shapes and natural frequencies thus generated are used as input for the identification process.

Modal analyses are also performed for cyclic sectors with nominal geometry and geometries perturbed along POD features to generate the basis vectors ($\Phi_0$, $\Phi_1$, and $\Phi_2$), mass and stiffness matrices ($M_t$ and $K_t$), and the gradients of mass and stiffness matrices along POD features ($\dfrac{\partial \delta M}{\partial \xi_s}$, $\dfrac{\partial \delta K}{\partial \xi_s}$, $\dfrac{\partial^2 \delta M}{\partial \xi_s \partial \xi_t}$ and $\dfrac{\partial^2 \delta K}{\partial \xi_s \partial \xi_t}$, $s, t = 1,2$). Mistuning identification is performed using the least square error solution of Equation (3.6.28) with data for 25–48 modes. Results are presented in Figures 3.6.1 (mistuning parameters for thickness) and 3.6.2 (mistuning parameters for blade surface inclination). The estimation error is calculated by taking the difference between the actual and estimated mistuning parameter values. As observed from the figures, the identification technique can estimate the values of mistuning parameters accurately, with mean and standard deviation of error ($\mu$, $\sigma$) for the first and second POD feature being (−0.3689e–3, 0.0012) and (0.2140e–3, 0.0007), respectively.

In the previous analysis, the nominal tuned geometry is selected as the average geometry. In practical situations, the mistuning parameters are unknown, therefore the true mean geometry is also unknown. Hence, the ideal true mean geometry can only be approximated by an available tuned nominal geometry. Obviously under such approximation the mean of the deviations of mistuning parameters from nominal values will not be equal to zero. To simulate this situation, the nominal tuned geometry is created with the blade thickness increased by 1 percent of the actual blade thickness. Note the difference in thickness of approximate nominal geometry and ideal nominal geometry (1 percent) is comparable to the perturbations in thicknesses of the actual blades (standard deviation 1.7 percent), hence this case evaluates the accuracy of the estimation technique under practical conditions. Modal analyses similar to previous case are again performed to generate the basis vectors ($\Phi_0$, $\Phi_1$, and $\Phi_2$), mass and stiffness matrices ($M_t$ and $K_t$), and the gradients

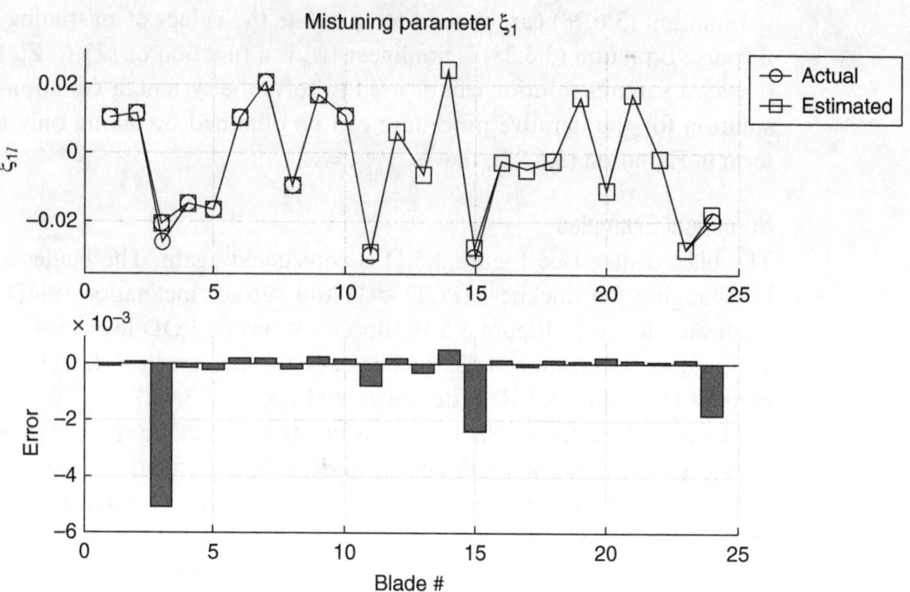

**Figure 3.6.1.**  Estimated mistuning parameter $\xi_1$ with true mean geometry (Bhartiya and Sinha, 2014).

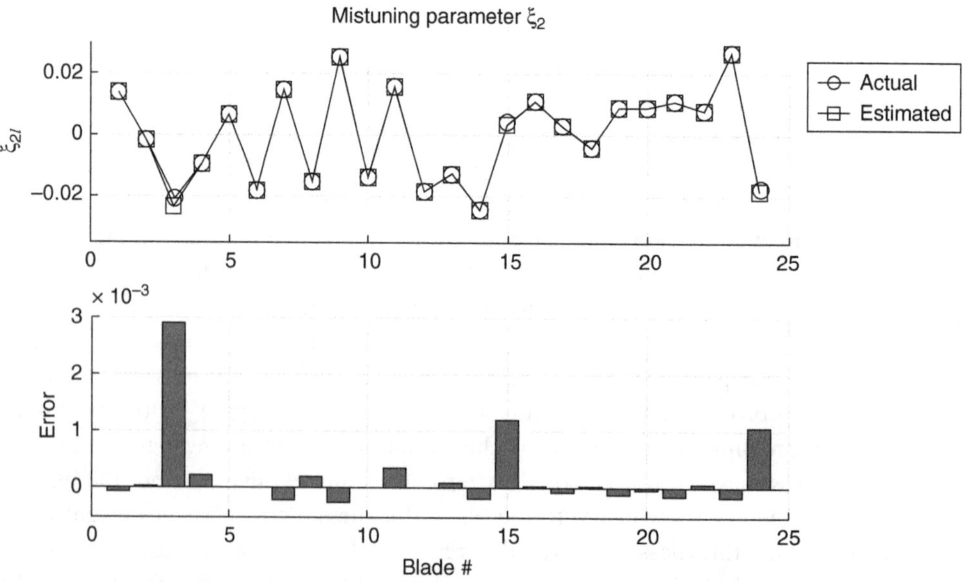

**Figure 3.6.2.**  Estimated mistuning parameter $\xi_2$ with true mean geometry (Bhartiya and Sinha, 2014).

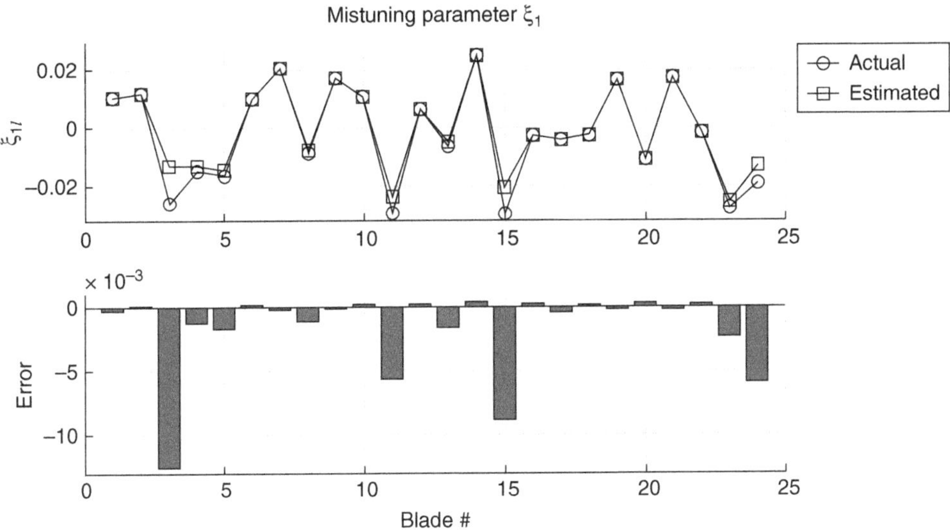


**Figure 3.6.3.** Estimated mistuning parameter $\xi_1$ estimated with approximate mean geometry (Bhartiya and Sinha, 2014).

of mass and stiffness matrices. Results from the mistuning identification algorithm are presented in Figures 3.6.3 and 3.6.4.

As observed from Figures 3.6.3 and 3.6.4, mistuning parameters estimated with approximate mean geometry are close to true values, with the mean values and standard deviations $(\mu, \sigma)$ of errors for POD1 and POD2 parameters being (−0.0016, 0.0033) and (0.0009, 0.0018), respectively. Because the mode shapes from the nominal and POD perturbed geometries are used to form the bases of MMDA, any error arising from the approximation of the average tuned geometry is getting compensated by the use of mode shapes from geometries perturbed along POD features.

## 3.7 Reduced-Order Model for Multistage Bladed Rotors

Compressor and turbine sections have many stages of bladed rotors. In the previous sections, each stage of bladed rotor was treated as isolated from other stages. But, recently, many papers have investigated the effects of structural coupling between rotor stages. Song, Castanier, and Pierre (2005) have developed a reduced-order model of a multistage rotor based on component mode synthesis. Laxalde, Thouverez, and Lombard (2007) have used sector analysis to find mode shapes of a multistage rotor in a subspace generated by the modes of individual disks. Sternchuss and Balmes (2007) have tried to overcome the necessity of compatible mesh at the disk interfaces by extending classical substructuring technique in cyclic

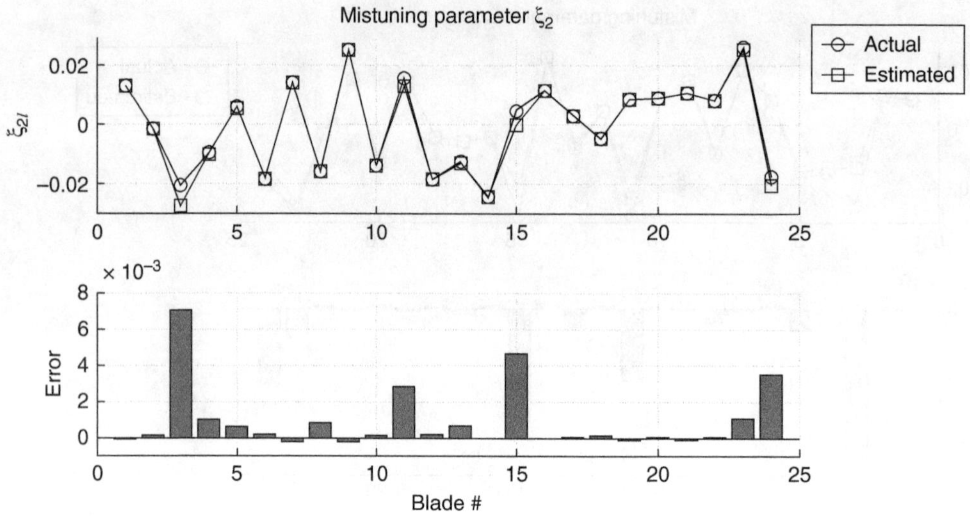

**Figure 3.6.4.**  Estimated mistuning parameter $\xi_2$ estimated with approximate mean geometry (Bhartiya and Sinha, 2014).

symmetry to compute mono-harmonic eigenvectors. It should be noted that papers by Song et al. (2005), Laxalde et al. (2007), and Sternchuss and Balmes (2007) do not deal with mistuning. Bladh, Castanier, and Pierre (2003) have examined the vibration of a two-stage mistuned rotor through a full (360 degree) finite element model. They have found cases in which where multistage analyses may be required.

Generally, each rotor stage does not have same number of blades. In this case, cyclic symmetry is lost even when all blades are identical (perfectly tuned) on each rotor stage, and sector analysis cannot be performed. The complication of vibration analysis is further enhanced in the presence of mistuning of blades. Using the idea behind MMDA, an accurate reduced-order model of a mistuned multistage rotor has been developed by Bhartiya and Sinha (2012).

### 3.7.1     Tuned Two-Stage Rotor

The free undamped vibration of a tuned two-stage rotor (see Figure 3.7.1) can be represented as

$$M_t\ddot{\mathbf{x}} + K_t\mathbf{x} = 0 \qquad (3.7.1)$$

where $M_t$ and $K_t$ are the mass and stiffness matrix of the tuned system with each blade having the average geometry and

$$\mathbf{x} = \begin{bmatrix} \mathbf{x}_L \\ \mathbf{x}_R \end{bmatrix} \qquad (3.7.2)$$

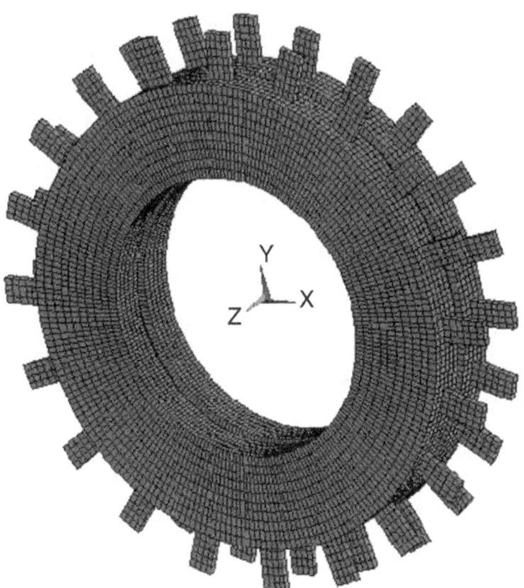

**Figure 3.7.1.** A two-stage rotor (Bhartiya and Sinha, 2012).

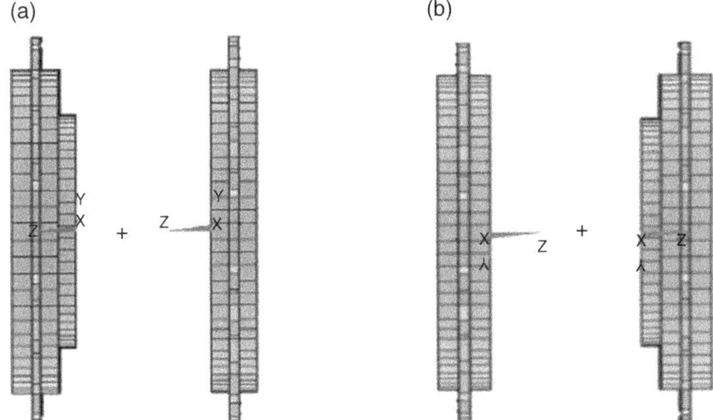


**Figure 3.7.2.** a. Connecting ring attached to left disk (Bhartiya and Sinha, 2012), b. Connecting ring attached to right disk.

where $\mathbf{x}_L$ and $\mathbf{x}_R$ are displacement vectors associated with left and right rotor stages, respectively. To develop a reduced-order model, basis vectors have been chosen as tuned modes of each bladed disk with connecting ring having both free and fixed ends (see Figure 3.7.2), that is,

$$\mathbf{x} = \Phi \mathbf{y} \tag{3.7.3}$$

where

$$\Phi = \begin{bmatrix} \phi_1 & \phi_{1f} & 0 & 0 \\ 0 & 0 & \phi_2 & \phi_{2f} \end{bmatrix} \qquad (3.7.4)$$

$\phi_1$: $r$ tuned modes of Disk 1 with the connecting ring having free end.

$\phi_{1f}$: $r$ tuned modes of Disk 1 with the connecting ring having fixed end.

$\phi_2$: $q$ tuned modes of Disk 2 with the connecting ring having free end.

and

$\phi_{2f}$: $q$ tuned modes of Disk 2 with the connecting ring having fixed end.

Substituting Equation (3.7.3) into Equation (3.7.1) and premultiplying by $\Phi^H$, the reduced-order model is obtained as follows:

$$M_r \ddot{\mathbf{y}} + K_r \mathbf{y} = 0 \qquad (3.7.5)$$

where

$$M_r = \Phi^H M_t \Phi \quad \text{and} \quad K_r = \Phi^H K_t \Phi \qquad (3.7.6)$$

Here, $K_t$ and $M_t$ are obtained by generating the mesh of the full 360-degree finite element model (FEM) and the modal vectors in the matrix $\Phi$ are obtained from modal analyses of finite element sectors. It should be noted that the modal analysis of the full 360-degree FEM is not required.

### Numerical Example

The FEM of a two-stage rotor is constructed. Numbers of blades on left and right disks are taken to be twenty-four and twenty, respectively. The natural frequencies and mode shapes are computed from the reduced-order model with $r = q = 120$, that is, the reduced-order model has the order of 480.

Natural frequencies and mode shapes are also computed from the full (360 degree) FEM of the two-stage rotor. Figure 3.7.3 shows the error in the natural frequency ($Freq_{ROM}$) predicted by the reduced-order model as a percentage of the frequency of the tuned system ($Freq_{Tnd}$) with average geometry estimated through 360-degree FEM analysis, that is, $Error(\%) = \dfrac{Freq_{ROM} - Freq_{Tnd}}{Freq_{Tnd}} * 100$, for the first 120 modes. As observed from the figure, the maximum error is 0.25 percent. The mean error is 0.0217 percent whereas the standard deviation of error is 0.0426 percent. Hence it can clearly be said that the reduced-order model provides almost exact natural frequencies of the system.

Next, the differences between mode shapes from the reduced-order model and full 360-degree FEM are also examined, for example, see Figure 3.7.4 in which an index $i$ of a modal vector represents its $i^{\text{th}}$ element. The modal vectors are scaled so that the maximum value of an element in a modal vector is 1. As observed from Figure 3.7.4, errors in the mode shapes predicted by the reduced-order model are very small (maximum value of error is 0.1 percent) and it can be again said that

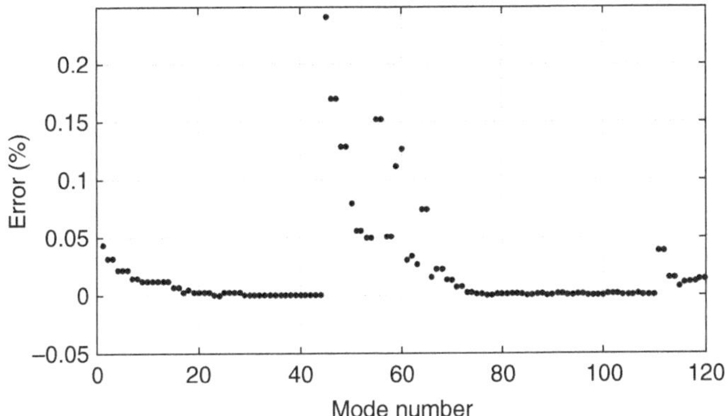

**Figure 3.7.3.**   Error (%) in frequency estimated through the reduced-order model (tuned two-stage rotor, $r = q = 120$) (Bhartiya and Sinha, 2012).

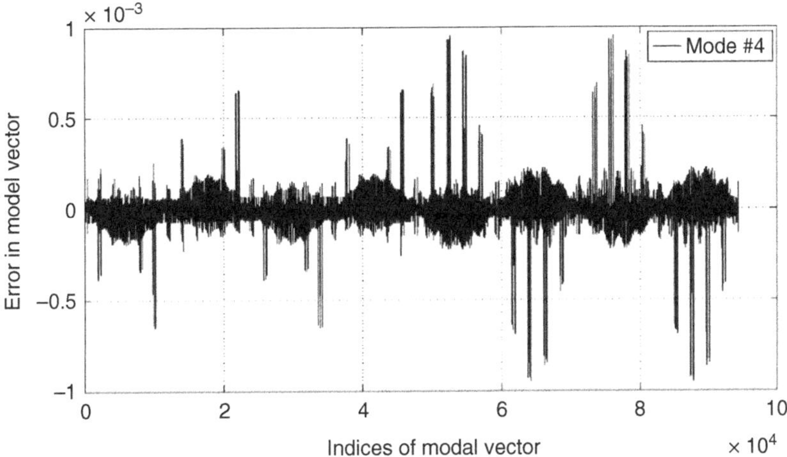


**Figure 3.7.4.**   Difference between modal vectors (mode number 4) from reduced-order model, and full rotor ANSYS analysis (tuned two-stage rotor, $r = q = 120$), Bhartiya and Sinha (2012).

the mode shapes are almost exact. This conclusion has also been verified by MAC (Allemang, 2003). The peaks in Figure 3.7.4 occur at the indices of the mode shape associated with the blades, which have much higher amplitudes compared to disks.

### 3.7.2    Mistuned Two-Stage Rotor

The reduced-order model for a two-stage rotor can be extended to include mistuning effects by including the tuned modes of the system with blades having perturbed geometry along a POD feature, that is,

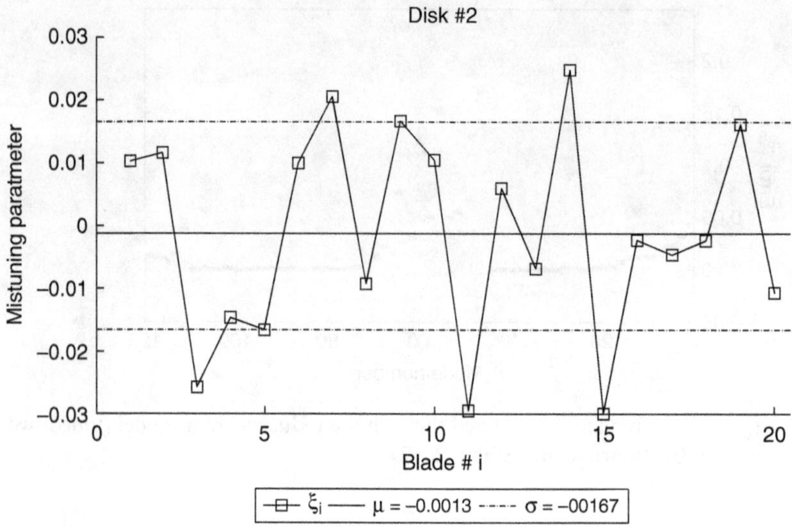


**Figure 3.7.5.**   Mistuning pattern for disk # 2 (Bhartiya and Sinha, 2012).

$$\Phi = \begin{bmatrix} \phi_1 & \phi_{1f} & \phi_{11} & . & . & \phi_{1p} \\ 0 & 0 & 0 & . & . & 0 \\ & & 0 & 0 & 0 & . & . & 0 \\ \phi_2 & \phi_{2f} & \phi_{21} & . & . & \phi_{2p} \end{bmatrix} \quad \cdots\cdots$$

(3.7.7)

where $\phi_{i\ell}$ are tuned modes of i[th] disk with blades having perturbed geometry along POD feature # $\ell$ (Sinha et al., 2008); $\ell = 1, 2, \ldots, p$.

## Numerical Example

The two-stage rotor considered in the previous section is used here again. The mistuning is introduced by varying the thickness of each blade. The thickness of the i[th] blade can be represented as:

$$t_i = t_0 (1 + \xi_i)$$

(3.7.8)

where $t_0$ is the thickness of the tuned blade on each disk. And, $\xi$ is the random variable with zero mean representing the only POD feature and $\xi_i$ is its value for the blade # $i$. Based upon a set of random values of $\xi$ from the Matlab routine "randn," a mistuning pattern is created for each disk (see Figure 3.5.3 for bladed disk # 1 and Figure 3.7.5 for bladed disk # 2). The standard deviation ($\sigma$) of $\xi$ is about 1.7 percent. Note that mean ($\mu$) is not exactly zero because of a finite number of random variables.

Figure 3.7.6 shows deviations in natural frequencies from their tuned values, which are calculated by the reduced-order model and the full (360 degree) ANSYS

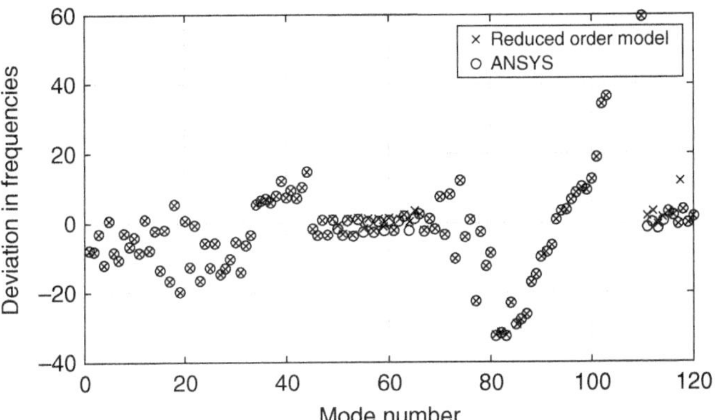

**Figure 3.7.6.**  Deviations in natural frequencies of a mistuned two-stage rotor ($r = q = 120$) (Bhartiya and Sinha, 2012).

analysis as well. As observed in Figure 3.7.6, the natural frequencies predicted by the reduced-order model are almost an exact match to those calculated by ANSYS. Some disagreements in the higher mode frequencies are expected because only 120 modes are used as a basis vectors for each $\Phi$ in Equation (3.7.7).

The percent error in the deviation from a tuned two-stage frequency ($Freq_{Tnd}$) is calculated as the ratio of the error in deviation calculated from the reduced-order model ($Dev_{ROM} - Dev_{Act}$) to the actual deviation ($Dev_{Act}$), that is,

$$Error D(\%) = \frac{Dev_{ROM} - Dev_{Act}}{Dev_{Act}} * 100 \tag{3.7.9}$$

where

$$Dev_{ROM} = Freq_{ROM} - Freq_{Tnd} \tag{3.7.10}$$

and

$$Dev_{Act} = Freq_{Act} - Freq_{Tnd} \tag{3.7.11}$$

As shown by Bhartiya and Sinha (2012), the percent errors in deviations are very small in spite of a strict criterion applied for the calculation of percent errors.

Figure 3.7.7 shows errors in modal vectors corresponding to mode 3 of the mistuned system. The difference between the modal vectors from the reduced-order model and ANSYS is almost zero. This conclusion has also been verified by MAC (Allemang, 2003).

### Forced Response with Proportional Damping

The steady-state harmonic response of the system can be calculated using mode superposition method. The system is excited by a harmonic force applied to Disk

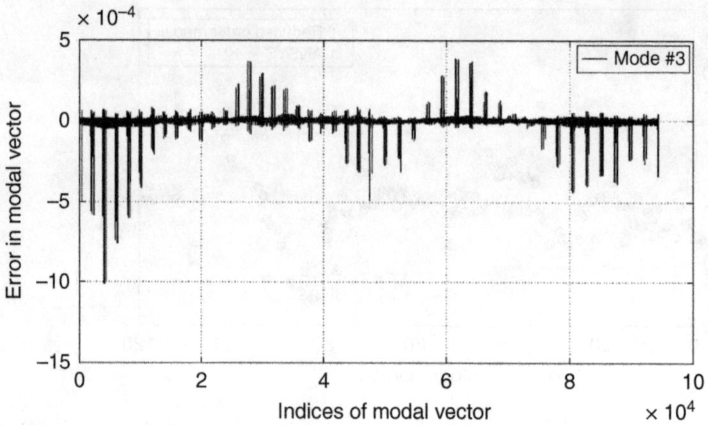

**Figure 3.7.7.**  Difference between mistuned modal vectors (mode 3) from reduced-order model and full ANSYS analysis ($r = q = 120$) (Bhartiya and Sinha, 2012).

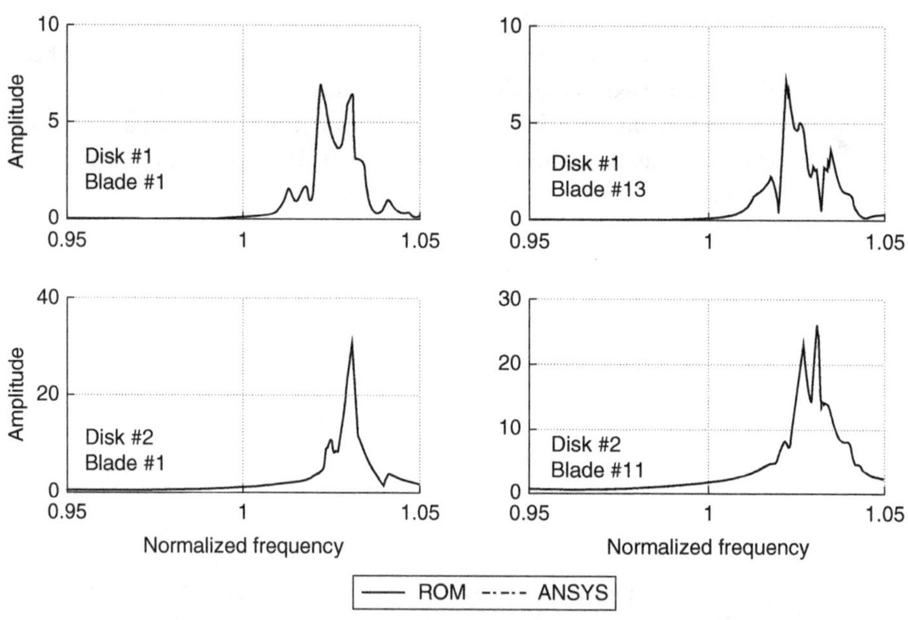

**Figure 3.7.8.**  Blade tip amplitude of disk 1 (top) and disk 2 (bottom) as a function of excitation frequencies ($r = q = 120$) (Bhartiya and Sinha, 2012).

2. The spatial distribution of the force corresponds to the second nodal diameter. No force is applied to Disk 1. Amplitudes of steady-state responses are obtained from the reduced-order model using the modal superposition method for excitation frequencies within 5 percent of the natural frequency corresponding to second nodal diameter of tuned single stage Disk 2. The damping ratio in each mistuned mode is taken to be 0.001. Figure 3.7.8 shows the tip amplitudes for Disk 1 (blade 1

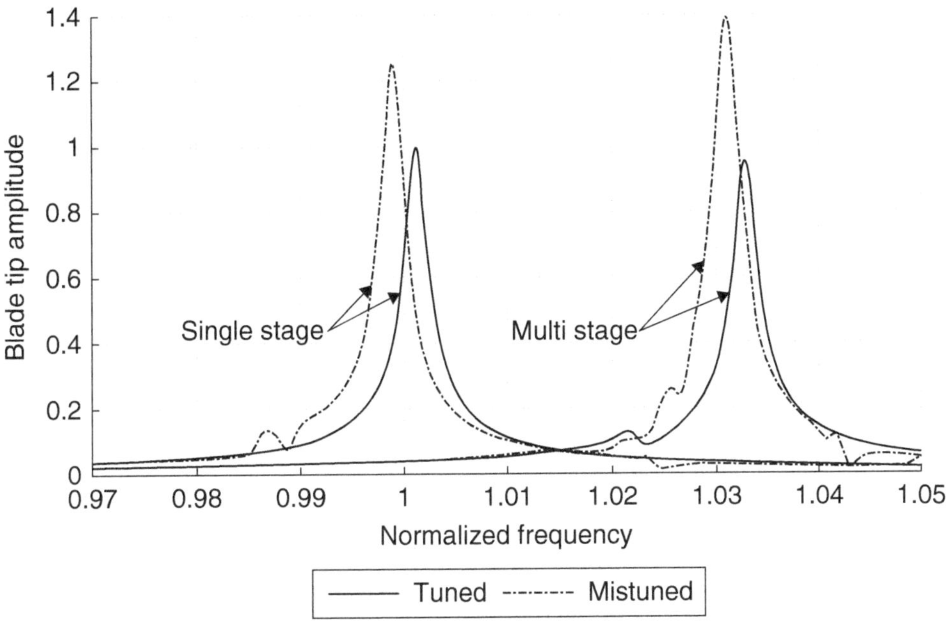



**Figure 3.7.9.** Forced responses of single-stage and multistage rotors (engine order 2, mean forcing frequency = 4102 Hz.) ($r = q = 120$) (Bhartiya and Sinha, 2012).

and blade 13) and Disk 2 (blade 1 and blade 11). These results match almost exactly with those from ANSYS analysis of full (360 degree) model.

Figure 3.7.9 represents the comparison between the forced responses of single-stage and multiple-stage rotors. The maximum amplitude of a mistuned two-stage rotor is higher than that of a mistuned single-stage rotor in this case.

# Appendix A    Fundamentals of Vibration Analysis of a Multidegree of Freedom System

## Eigenvalues and Eigenvectors (Sinha 2010)

Free vibration of an undamped multidegree of freedom system is given by

$$M\ddot{\mathbf{x}} + K\mathbf{x} = \mathbf{0} \tag{A.1}$$

where $M$ and $K$ are symmetric mass and stiffness matrices, respectively.
Let

$$\mathbf{x}(t) = \mathbf{a}\sin(\omega t + \phi) \tag{A.2}$$

where $n \times 1$ vector $\mathbf{a}$, the frequency $\omega$, and the phase $\phi$ are to be determined.
Differentiating (A.2) twice with respect to time,

$$\ddot{\mathbf{x}} = -\omega^2 \mathbf{a}\sin(\omega t + \phi) \tag{A.3}$$

Substituting (A.2) and (A.3) into (A.1),

$$(K - \omega^2 M)\mathbf{a} = \mathbf{0} \tag{A.4}$$

For a nonzero or nontrivial solution of $\mathbf{a}$,

$$\det(K - \omega^2 M) = 0 \tag{A.5}$$

which will be a polynomial equation of degree $n$ in $\omega^2$. Equation (A.4) can also be written as

$$K\mathbf{a} = \omega^2 M\mathbf{a} \tag{A.6}$$

Equation (A.6) suggests that $\omega^2$ and $\mathbf{a}$ are generalized eigenvalues and eigenvectors of the stiffness matrix $K$ with respect to the mass matrix $M$. The Matlab command for computation of generalized eigenvalues and eigenvectors is $eig(K, M)$. The formulation of the generalized eigenvalue/eigenvector problem is convenient for many degrees of freedom because the inverse of mass matrix is not required.

## Orthoganality of Eigenvectors for Symmetric Mass and Symmetric Stiffness Matrices (Sinha 2010)

Let $\omega_i^2$ and $\mathbf{v}_i$ be eigenvalue and eigenvector pair where $i = 1, 2, \ldots, n$. Then,

$$K\mathbf{v}_i = \omega_i^2 M\mathbf{v}_i \tag{A.7}$$

and

$$K\mathbf{v}_j = \omega_j^2 M\mathbf{v}_j \tag{A.8}$$

Premultiplying both sides of (A.7) by $\mathbf{v}_j^T$,

$$\mathbf{v}_j^T K\mathbf{v}_i = \omega_i^2 \mathbf{v}_j^T M\mathbf{v}_i \tag{A.9}$$

Premultiplying both sides of (A.8) by $\mathbf{v}_i^T$ and then taking transpose,

$$(\mathbf{v}_i^T K\mathbf{v}_j)^T = \omega_j^2 (\mathbf{v}_i^T M\mathbf{v}_j)^T \tag{A.10}$$

or

$$\mathbf{v}_j^T K^T \mathbf{v}_i = \omega_j^2 \mathbf{v}_j^T M^T \mathbf{v}_i \tag{A.11}$$

For symmetric mass and symmetric stiffness matrices,

$$K = K^T \tag{A.12}$$

and

$$M = M^T \tag{A.13}$$

Using Equations (A.12) and (A.13), Equation (A.11) yields

$$\mathbf{v}_j^T K\mathbf{v}_i = \omega_j^2 \mathbf{v}_j^T M\mathbf{v}_i \tag{A.14}$$

Substituting (A.9) into (A.14),

$$(\omega_i^2 - \omega_j^2)\mathbf{v}_j^T M\mathbf{v}_i = 0 \tag{A.15}$$

As a result,

$$\mathbf{v}_j^T M\mathbf{v}_i = 0 \quad \text{for} \quad \omega_i \neq \omega_j \tag{A.16}$$

From (A.9) and (A.16),

$$\mathbf{v}_j^T K\mathbf{v}_i = 0 \quad \text{for} \quad \omega_i \neq \omega_j \tag{A.17}$$

Usually, each eigenvector is scaled such that

$$\mathbf{v}_i^T M\mathbf{v}_i = 1; \quad i = 1, 2, \ldots, n \tag{A.18}$$

In this case, from Equation (A.11),

$$\mathbf{v}_i^T K \mathbf{v}_i = \omega_i^2; \quad i = 1, 2, \ldots, n \tag{A.19}$$

Define a modal matrix $V$ as follows:

$$V = [\mathbf{v}_1 \quad \mathbf{v}_2 \quad \cdots \quad \mathbf{v}_{n-1} \quad \mathbf{v}_n] \tag{A.20}$$

Then, Equations (A.17) through (A.19) are expressed as

$$V^T M V = I_n \tag{A.21}$$

and

$$V^T K V = \Lambda \tag{A.22}$$

where

$$\Lambda = \begin{bmatrix} \omega_1^2 & 0 & \cdots & 0 & 0 \\ 0 & \omega_2^2 & \cdots & 0 & 0 \\ \vdots & \vdots & \ddots & \vdots & \vdots \\ 0 & 0 & \cdots & \omega_{n-1}^2 & 0 \\ 0 & 0 & \cdots & 0 & \omega_n^2 \end{bmatrix} \tag{A.23}$$

**Note:**
The derivations (A.21) and (A.22) are shown only for nonrepeated natural frequencies. However, it is also possible to diagonalize when some of the natural frequencies are repeated.

## Modal Decomposition

Forced vibration of a damped multidegree of freedom system is given by

$$M\ddot{\mathbf{x}} + C\dot{\mathbf{x}} + K\mathbf{x} = \mathbf{f}(t) \tag{A.24}$$

where $C$ and $\mathbf{f}(t)$ are damping and external force vector, respectively.

In general, the response $\mathbf{x}(t)$ is a linear combination of modal vectors $\mathbf{v}_i$; $i = 1, 2, \ldots, n$, that is,

$$\mathbf{x}(t) = \mathbf{v}_1 y_1(t) + \mathbf{v}_2 y_2(t) + \cdots + \mathbf{v}_n y_n(t) \tag{A.25}$$

where $y_i(t)$ is the coefficient of the modal vectors $\mathbf{v}_i$, $i = 1, 2, \ldots, n$. Equation (A.25) can be represented in a compact form as follows:

$$\mathbf{x}(t) = V\mathbf{y}(t) \tag{A.26}$$

where the matrix $V$ is defined by Equation (A.20), and the vector $\mathbf{y}(t)$ is defined as

$$\mathbf{y}(t) = [\,y_1 \quad y_2 \quad \cdots \quad y_{n-1} \quad y_n\,]^T \tag{A.27}$$

Substituting (A.26) into (A.24), and premultiplying by $V^T$,

$$V^T M V \ddot{\mathbf{y}} + V^T C V \dot{\mathbf{y}} + V^T K V \mathbf{y} = V^T \mathbf{f}(t) \tag{A.28}$$

Equation (A.28) are often called *modal equations* as they are in terms of modal components $y_i$; $i = 1, 2, \ldots, n$. Matrices $V^T M V$ and $V^T K V$ are diagonal equations (see (A.21) and (A.22)), but there is no guarantee that $V^T C V$ is diagonal. Two special cases of damping resulting in decoupled modal equations are considered as follows.

**Case I: Undamped System ($C = 0$)**
Substituting Equations (A.21) and (A.22) into Equation (A.28),

$$\ddot{\mathbf{y}} + \Lambda \mathbf{y} = V^T \mathbf{f}(t) \tag{A.29}$$

or

$$\ddot{y}_i + \omega_i^2 y_i = \mathbf{v}_i^T \mathbf{f}(t); \; i = 1, 2, \ldots, n \tag{A.30}$$

Here, modal equations are decoupled and each modal equation (A.30) can be viewed as an equivalent undamped single degree of freedom system subjected to the force $\mathbf{v}_i^T \mathbf{f}(t)$. The quantity $\mathbf{v}_i^T \mathbf{f}(t)$ is also known as the *modal force*.

**Case II: Proportional or Rayleigh Damping**
Assume that the damping matrix has the following form:

$$C = \alpha M + \beta K \tag{A.31}$$

where $\alpha$ and $\beta$ are constants. This form of damping is known as *proportional* or *Rayleigh* damping. Substituting (A.21), (A.22), and (A.31) into (A.28),

$$\ddot{\mathbf{y}} + (\alpha I + \beta \Lambda)\dot{\mathbf{y}} + \Lambda \mathbf{y} = V^T \mathbf{f}(t) \tag{A.32}$$

or

$$\ddot{y}_i + (\alpha + \beta \omega_i^2)\dot{y}_i + \omega_i^2 y_i = \mathbf{v}_i^T \mathbf{f}(t); \; i = 1, 2, \ldots, n \tag{A.33}$$

Again, modal equations are decoupled and each modal equation (A.33) can be viewed as an equivalent damped single degree of freedom system subjected to the modal force $\mathbf{v}_i^T \mathbf{f}(t)$.

# Appendix B  Eigenvalues and Eigenvectors of a Circulant Matrix

Consider the following $n \times n$ circular matrix:

$$V = \begin{bmatrix} v_1 & v_2 & 0 & 0 & \cdots & 0 & v_2 \\ v_2 & v_1 & v_2 & 0 & \cdots & 0 & 0 \\ 0 & v_2 & v_1 & v_2 & \cdots & 0 & 0 \\ \cdots & & & & & & \cdots \\ v_2 & 0 & 0 & 0 & \cdots & v_2 & v_1 \end{bmatrix} \tag{B.1}$$

Further, consider $n^{th}$ root of unity, that is,

$$(1)^{\frac{1}{n}} = (e^{j2\pi\ell})^{\frac{1}{n}} = e^{j\phi\ell}; \quad \ell = 0,1,2,\ldots,(n-1) \tag{B.2}$$

where $j = \sqrt{-1}$ and

$$\phi = \frac{2\pi}{n} \tag{B.3}$$

For each of the $n^{th}$ root of unity, the following $n \times 1$ vector is constructed:

$$\mathbf{p}_\ell = [1 \quad e^{j\phi\ell} \quad e^{j2\phi\ell} \quad \cdots \quad e^{j(n-1)\phi\ell}]^T; \quad \ell = 0,1,2,\ldots,(n-1) \tag{B.4}$$

Then, it can be shown that

$$V\,\mathbf{p}_\ell = \lambda_\ell\,\mathbf{p}_\ell \tag{B.5}$$

where

$$\lambda_\ell = v_1 + v_2 e^{j\phi\ell} + v_2 e^{j(n-1)\phi\ell} = v_1 + v_2 e^{j\phi\ell} + v_2 e^{-j\phi\ell} = v_1 + 2v_2 \cos\phi\ell \tag{B.6}$$

The proof of result (B.5) is easily seen by finding $m^{th}$ element of $V\,\mathbf{p}_\ell$:

$$v_2 e^{j(m-2)\phi\ell} + v_1 e^{j(m-1)\phi\ell} + v_2 e^{jm\phi\ell} = (v_1 + v_2 e^{j\phi\ell} + v_2 e^{-j\phi\ell})e^{j(m-1)\phi\ell} \tag{B.7}$$

Relationship (B.5) suggests that $\lambda_\ell$ and $\mathbf{p}_\ell$ are eigenvalues and corresponding eigenvectors of the matrix $V$. Because the determinant of a matrix equals product of its eigenvalues:

$$\det V = \prod_{\ell=0}^{n-1} \lambda_\ell \tag{B.8}$$

Next, consider the following eigenvector:

$$\mathbf{p}_{n-\ell} = [1 \quad e^{j\phi(n-\ell)} \quad e^{j2\phi(n-\ell)} \quad \cdots \quad e^{j(n-1)\phi(n-\ell)}]^T \tag{B.9}$$

Because $n\phi = 2\pi$,

$$\mathbf{p}_{n-\ell} = [1 \quad e^{-j\phi\ell} \quad e^{-j2\phi\ell} \quad \cdots \quad e^{-j(n-1)\ell)}]^T \tag{B.10}$$

From Equation (B.5), the eigenvalue $\lambda_{n-\ell}$ corresponding to the eigenvector $\mathbf{p}_{n-\ell}$ is given by

$$\lambda_{n-\ell} = v_1 + v_2 e^{-j\phi\ell} + v_2 e^{-j(n-1)\phi\ell} = v_1 + v_2 e^{-j\phi\ell} + v_2 e^{j\phi\ell} = v_1 + 2v_2 \cos(\phi\ell) \tag{B.11}$$

Comparing Equation (B.5) and Equation (B.11),

$$\lambda_{n-\ell} = \lambda_\ell; \; \ell > 0 \tag{B.12}$$

Further,

$$\mathbf{p}_{n-\ell}^H \mathbf{p}_\ell = \begin{bmatrix} 1 \\ e^{-j\ell\phi} \\ e^{-j2\ell\phi} \\ \vdots \\ e^{-j(n-1)\ell\phi} \end{bmatrix}^H \begin{bmatrix} 1 \\ e^{j\ell\phi} \\ e^{j2\ell\phi} \\ \vdots \\ e^{j(n-1)\ell\phi} \end{bmatrix} = 1 + e^{j2\ell\phi} + e^{j4\ell\phi} + \cdots + e^{j2(n-1)\ell\phi} \tag{B.13}$$

where $\mathbf{p}_{n-\ell}^H$ is the complex conjugate transpose of $\mathbf{p}_{n-\ell}$. Because the right-hand side of Equation (B.13) is a geometric progression,

$$\mathbf{p}_{n-\ell}^H \mathbf{p}_\ell = 1 + e^{j2\ell\phi} + e^{j4\ell\phi} + \cdots + e^{j2(n-1)\ell\phi} = \frac{1 - e^{j2\phi n}}{1 - e^{j2\phi}} \tag{B.14}$$

Therefore, using Equation (B.3),

$$\mathbf{p}_{n-\ell}^H \mathbf{p}_\ell = \frac{1 - e^{j4\pi}}{1 - e^{j2\varphi}} = 0 \tag{B.15}$$

In other words, there are eigenvalues that are repeated twice. Further, there are two independent and orthogonal eigenvectors $\mathbf{p}_{n-\ell}$ and $\mathbf{p}_\ell$ corresponding to a repeated eigenvalue because of Equation (B.15). Also, constant phase angle between two successive elements for $\mathbf{p}_{n-\ell}$ is negative of that for $\mathbf{p}_\ell$, Equation (B.4) and Equation (B.9).

Further, from Equations (B.14) and (B.15),

$$1 + \cos(2\phi\ell) + \cos(4\phi\ell) + \cdots + \cos(2(n-1)\phi\ell) = 0 \tag{B.16}$$

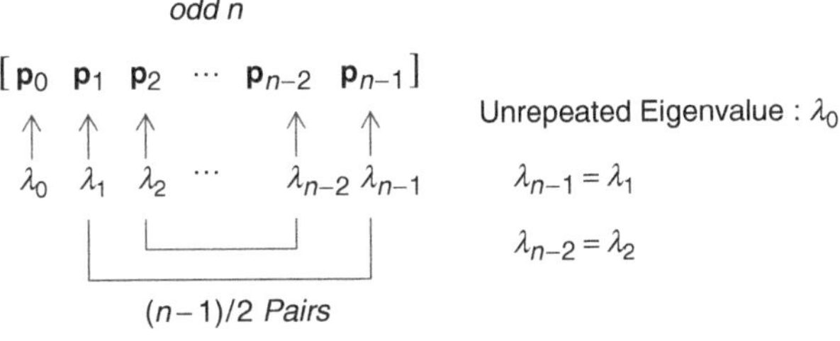



**Figure B.1.**   Eigenvalues and eigenvectors for odd $n$.

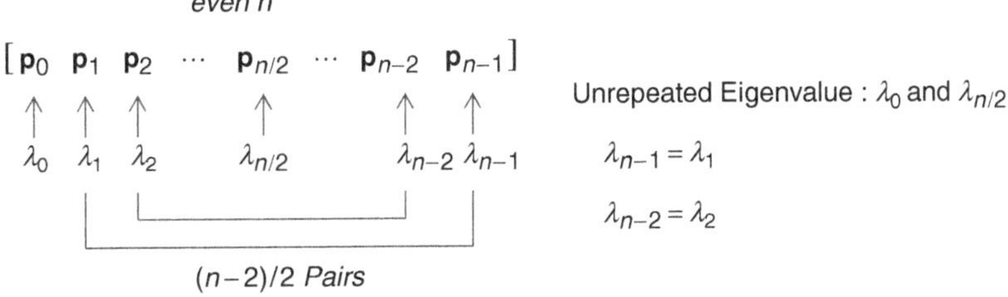



**Figure B.2.**   Eigenvalues and eigenvectors for even $n$.

$$\sin(2\phi\ell) + \sin(4\phi\ell) + \cdots + \sin(2(n-1)\phi\ell) = 0 \tag{B.17}$$

For odd $n$, there are $(n-1)/2$ repeated eigenvalues (see Figure B.1). There is only one unrepeated eigenvalue corresponding to $\ell = 0$ with the eigenvector

$$\mathbf{p}_0 = [1 \quad 1 \quad 1 \quad \cdots \quad 1]^T \tag{B.18}$$

For even $n$, there are $(n-2)/2$ repeated eigenvalues (see Figure B.2). There are two unrepeated eigenvalues corresponding to $\ell = 0$ and $\ell = n/2$ with eigenvectors $\mathbf{p}_0$, Equation (B.16), and

$$\mathbf{p}_{n/2} = [1 \quad -1 \quad 1 \quad \cdots \quad -1]^T \tag{B.19}$$

Also, the constant phase angle between two successive elements for $\mathbf{p}_{n/2}$ is $180^0$, which is not possible for odd $n$.

Lastly, the matrix $V$ is symmetric. Therefore, eigenvalues corresponding to distinct eigenvalues are orthogonal (Strang, 1988). In other words, all eigenvectors given by (B.4) are orthogonal. Let

$$\Phi_v = \begin{bmatrix} \mathbf{p}_1 & \mathbf{p}_2 & \cdots & \mathbf{p}_{n-1} & \mathbf{p}_n \end{bmatrix} \tag{B.20}$$

Then,

$$\Phi_v^{-1} = \frac{1}{n}\Phi_v^H \tag{B.21}$$

where $\Phi_v^H$ is the complex conjugate transpose of $\Phi_v$.

# References


Adamczyk, J. J., and Goldstein, M. E., 1978, "Unsteady Flow in Supersonic Cascade with Subsonic Leading-Edge Locus," *AIAA Journal*, Vol. 16, No. 12, December, pp. 1248–1254.

Allemang, R. J., 2003, "Modal Assurance Criterion – Twenty Years of Use and Abuse," *Sound and Vibration*, August, pp. 14–21.

Allen, P. B., and Kelnar, J., 1998, "Evolution of a Vibrational Wave Packet on a Disordered Chain," *American Journal of Physics*, Vol. 66, No. 6, pp. 497–506.

Anderson, P. W., 1958, "Absence of Diffusion in Certain Random Lattices," *Physical Review*, Vol. 109, No. 5, pp. 1492–1505.

Andrew, A. L., and Tan, R. C. E., 1998, "Computation of Derivatives of Repeated Eigenvalues and the Corresponding Eigenvectors of Symmetric Matrix Pencils," *SIAM Journal of Matrix Analysis and Applications*, Vol. 20, No. 1, pp. 78–100.

*ANSYS* Theory Manual, 2012, 12th ed., pp. 15–65 to 15–66.

Beck, J. A., 2010, "Stochastic Mistuning Simulation of Integrally Bladed Rotors Using Nominal and Non-nominal Component Mode Synthesis," MS Thesis, Wright State University, Dayton, OH.

Bhartiya, Y., and Sinha, A., 2011, "Reduced Order Model of a Bladed Rotor with Geometric Mistuning: Comparison between Modified Modal Domain Analysis and Frequency Mistuning Approach," GT2011-45391, presented at the ASME Turbo Expo, Vancouver, BC, Canada, June 6–11.

Bhartiya, Y., and Sinha, A., 2012, "Reduced Order Model of a Multistage Bladed Rotor with Geometric Mistuning via Modal Analyses of Finite Element Sectors," *ASME Journal of Turbomachinery*, Vol. 134, July, pp. 041001-1 to 041001-8.

Bhartiya, Y., and Sinha, A., 2013a, "Reduced Order Model of a Bladed Rotor with Geometric Mistuning via Estimated Deviations in Mass and Stiffness Matrices," *ASME Journal of Engineering for Gas Turbines and Power*, Vol. 135, May, pp. 052501-1 to 052501-8.

Bhartiya, Y., and Sinha, A., 2013b, "Reduced Order Modeling of a Bladed Rotor with Geometric Mistuning: Alternative Bases and Extremely Large Mistuning," *International Journal of Gas Turbine, Propulsion and Power System*, Vol. 5, No. 1, pp. 1–7.

Bhartiya, Y., and Sinha, A., 2014, "Geometric Mistuning Identification of Integrally Bladed Rotors Using Modified Modal Domain Analysis," *ASME Journal of Engineering for Gas Turbines and Power*, Vol. 136, December, pp. 122504-1 to 122504-8.

Bladh, R., Castanier, M. P., and Pierre, C., 2001, "Component-Mode-Based Reduced Order Modeling Techniques for Mistuned Bladed Disks – Part I: Theoretical Models, Part II: Application," *ASME Journal of Engineering for Gas Turbines and Power*, Vol. 123, January, pp. 89–108.

Bladh, R., Castanier, M. P., and Pierre, C., 2003, "Effects of Multistage Coupling and Disk Flexibility on Mistuned Bladed Disk Dynamics," *ASME Journal of Engineering for Gas Turbines and Power*, Vol. 125, pp. 121–130.

Brown, J. M., 2008, "Reduced Order Modeling Methods for Turbomachinery Design," PhD Dissertation, Wright State University, Dayton, OH.

Castanier, M. P., and Pierre, C., 1997, "Consideration on the Benefits of Intentional Blade Mistuning for the Forced Response of Turbomachinery Rotors," *Proceedings of the ASME Aerospace Division*, AD – 55, pp. 419–425.

Chan, Y.-J., and Ewins, D. J., 2011, "The Amplification of Vibration Response Levels of Mistuned Bladed Disks: Its Consequences and Its Distribution in Specific Situations," *ASME Journal of Engineering for Gas Turbines and Power*, Vol. 133, October, 102502-1 to 102502-8.

Chen, G., 2005, *Nanoscale Energy Transport and Conversion*, Oxford: Oxford University Press.

Choi, B-K., Lentz, J., Rivas-Guerra, A., and Mignolet, M. P., 2003, "Optimization of Intentional Mistuning Patterns for the Reduction of the Forced Response Effects of Unintentional Mistuning: Formulation and Assessment," *ASME Journal of Engineering for Gas Turbines and Power*, Vol. 125, pp. 131–140.

Cleland, A. N., 2003, *Foundations of Nanomechanics*, Berlin: Springer-Verlag.

Craig, R. R., 1981, *Structural Dynamics: An Introduction to Computer Methods*, New York: John Wiley and Sons.

Crawley, E. F., and Hall, K. C., 1985, "Optimization and Mechanisms of Mistuning in Cascades," *ASME Journal of Engineering for Gas Turbines and Power*, Vol. 107, April, pp. 418–426.

D'Amato, F., and Rotea, M., 2005, "LFTB: An Efficient Algorithm to Bound Linear Fractional Transformations," *Optimization and Engineering*, Vol. 6, pp. 177–201.

Dowell, E. H., 2014, *A Modern Course in Aeroelasticity*, Berlin: Springer.

Ewins, D. J., 1969, "The Effects of Detuning upon the Forced Vibrations of Bladed Disks," *Journal of Sound and Vibration*, Vol. 9, No. 1, pp. 65–79.

Fan, M., Tits, A., and Doyle, J. C., 1991, "Robustness in the Presence of Joint Parametric Uncertainty and Unmodeled Dynamics," Vol. 36, No. 1, pp. 25–38.

Feiner, D. M., and Griffin, J. H., 2002, "A Fundamental Model of Mistuning for a Single Family of Modes," *ASME Journal of Turbomachinery*, Vol. 124, No. 4, pp. 597–605.

Feiner, D.M., and Griffin, J. H., 2004a, "Mistuning Identification of Bladed Disks Using a Fundamental Mistuning Model – Part 1: Theory," *ASME Journal of Turbomachinery*, Vol. 126, No. 1, pp. 150–158.

Feiner, D.M., and Griffin, J. H., 2004b, "Mistuning Identification of Bladed Disks Using a Fundamental Mistuning Model – Part 2: Application," *ASME Journal of Turbomachinery*, Vol. 126, No. 1, pp. 159–165.

Fox, R. L., and Kapoor, M. P., 1968, "Rates of Change of Eigenvalues and Eigenvectors," *AIAA Journal*, Vol. 6, No. 12, pp. 2426–2429.

Ghanem, R. G., and Spanos, P. D., 1990, *Stochastic Finite Elements: A Spectral Approach*, New York: Springer-Verlag.

Jones, K., 2004, "Mistuning for Minimum Maximum Bladed Disk Forced Response," AIAA 2004–3755, 40th AIAA/ASME/SAE/ASEE Joint Propulsion Conference and Exhibit, Fort Lauderdale, FL, July.

Jones, K., and Cross, C., 2003, "On Antiresonance in the Forced Response of Mistuned Bladed Disks," *Shock and Vibration*, Vol. 10, pp. 135–146.

Kaza, K. R. V., and Kielb, R. E., 1982, "Flutter and Response of a Mistuned Cascade in Incompressible Flow," *AIAA Journal*, Vol. 20, No. 8, August, pp. 1120–1127.

Kenyon, J. A., and Griffin, J. H., 2003, "Experimental Demonstration of Maximum Mistuned Bladed Disk Forced Response," *ASME Journal of Turbomachinery*, Vol. 125, October, pp. 673–681.

Kenyon, J. A., Griffin, J. H., and Feiner, D. M., 2003, "Maximum Bladed Disk Forced Response from Distortion of a Structural Mode," *ASME Journal of Turbomachinery*, Vol. 125, April, pp. 353–363.

Kenyon, J. A., Griffin, J. H., and Kim, N. E., 2005, "Sensitivity of Tuned Bladed Disk Response to Frequency Veering," *ASME Journal of Engineering for Gas Turbines and Power*, Vol. 127, October, pp. 835–842.

Kielb, R. E., and Kaza, K. R.V., 1983, "Aeroelastic Characteristics of a Cascade of Mistuned Blades in Subsonic and Supersonic Flows," *ASME Journal of Vibration, Acoustics, Stress, and Reliability in Design*, Vol. 105, pp. 425–433.

Kittel, C., 1996, *Introduction to Solid State Physics*, 7th ed., New York: John Wiley and Sons.

Laxalde, D., Thouverez, F., and Lombard, J.-P., 2007, "Dynamical Analysis of Multi-Stage Cyclic Structures," *Mechanics Research Communication*, Vol. 34, pp. 379–384.

Lim, S. H., Pierre, C., and Castanier, M.P., 2004, "Mistuning Identification and Reduced-Order Model Updating for Bladed Disks on a Component Mode Mistuning Technique," 9th National Turbine Engine High Cycle Fatigue Conference, Pinehurst, North Carolina.

MATLAB, 1992, "MATLAB Reference Guide," The MathWorks, Inc., Natick, MA.

Mignolet, M. P., and Lin, C.-C., 1993, "The Combined Closed-Form Perturbation Approach to the Analysis of Mistuned Bladed Disks," *ASME Journal of Turbomachinery*, Vol. 115, October, pp. 771–780.

Mignolet, M. P., Delor, J. P., and Rivas-Guerra, A., 1999a, "Identification of Mistuning Characteristics of Bladed Disks from Free Response Data – Part I," *ASME Journal of Engineering for Gas Turbines and Power*, Vol. 123, pp. 395–403.

Mignolet, M. P., Delor, J. P., and Rivas-Guerra, A., 1999b, "Identification of Mistuning Characteristics of Bladed Disks from Free Response Data – Part II," *ASME Journal of Engineering for Gas Turbines and Power*, Vol. 123, pp. 404–411.

Mignolet, M. P., Hu, W., and Jadic, I., 2000, "On the Forced Response of Harmonically and Partially Mistuned Bladed Disks, Part I: Harmonic Mistuning, Part II: Partial Mistuning and Applications," *International Journal of Rotating Machinery*, Vol. 6, pp. 29–56.

Mignolet, M. P., Rivas-Guerra, A., and LaBorde, B., 1999, "Towards a Comprehensive Direct Prediction Strategy of the Effects of Mistuning on the Forced Response of Turbomachinery Blades," *Aircraft Engineering and Aerospace Technology*, Vol. 71, No. 5, pp. 462–469.

Nelson, R. B., 1976, "Simplified Calculation of Eigenvector Derivatives," *AIAA Journal*, Vol. 14, pp. 1201–1205.

Oppenheim, A. V., and Shaeffer, R. W., 1975, *Digital Signal Processing*, Englewood Cliffs, NJ: Prentice-Hall.

Papoulis, A., 1965, *Probability, Random Variables, and Stochastic Processes*, New York: McGraw-Hill.

Petrov, E. P., and Ewins, D. J., 2003, "Analysis of the Worst Mistuning Patterns in Bladed Disk Assemblies," *ASME Journal of Turbomachinery*, Vol. 125, October, pp. 623–631.

Phani, A. S., Woodhouse, J., and Fleck, N. A., 2006, "Wave Propagation in Two-Dimensional Periodic Lattices," *Journal of Acoustical Society of America*, Vol. 119, No. 4, pp. 1995–2005.

Prescott, J., 1946, *Applied Elasticity*, New York: Dover Publications.

Rao, J. S., 1991, *Turbomachine Blade Vibration*, New York: John Wiley and Sons.

Rivas-Guerra, A. J., and Mignolet, M. P., 2003, "Maximum Amplification of Blade Response Due to Mistuning: Localization and Mode Shape Aspects of the Worst Disks," *ASME Journal of Turbomachinery*, Vol. 125, July, pp. 442–454.

Rotea, M., and D' Amato, F., 2002, "New Tools for Analysis and Optimization of Mistuned Bladed Disks," AIAA 2002 – 4081, Proceedings of the 38th AIAA/ASME/SAE/ASEE Joint Propulsion Conference and Exhibit, July, Indianapolis, IN.

Shapiro, B., 1998, "Symmetry Approach to Extension of Flutter Boundaries via Mistuning," *AIAA Journal of Propulsion and Power*, Vol. 14, No. 3, pp. 354–366.

Shapiro, B., 1999, "Solving for Mistuned Forced Response by Symmetry," *AIAA Journal of Propulsion and Power*, Vol. 15, No. 2, pp. 310–325.

Sinha, A., 1986, "Calculating the Statistics of Forced Response of a Mistuned Bladed Disk Assembly," *AIAA Journal*, Vol. 24, No. 11, pp. 1797–1801.

Sinha, A., 1997, "Computing the Maximum Amplitude of a Mistuned Bladed Disk Assembly via Infinity Norm," *Proceedings of the ASME Aerospace Division*, AD – 55, pp. 427–432.

Sinha, A., 2005, "Statistical Distribution of the Peak Maximum Amplitude of the Forced Response of a Mistuned Bladed Disk," GT2005 – 69070, ASME Turbo Expo, Reno-Tahoe, NV.

Sinha, A., 2006a, "Computation of Eigenvalues and Eigenvectors a Mistuned Bladed Disk," GT2006 – 90087, ASME Turbo Expo, Barcelona, Spain.

Sinha, A., 2006b, "Statistical Distribution of the Peak Maximum Amplitude of the Forced Response of a Mistuned Bladed Disk: Some Investigations and Applications of Neural Networks," GT2006 – 90088, ASME Turbo Expo, Barcelona, Spain.

Sinha, A., 2006c, "Computation of the Statistics of Forced Response of a Mistuned Bladed Disk Assembly via Polynomial Chaos," *ASME Journal of Vibration and Acoustics*, Vol. 128, August, pp. 449–457.

Sinha, A., 2007, *Linear Systems: Optimal and Robust Control*, Boca Raton, FL: CRC Press (Taylor Francis Group).

Sinha, A., 2008a, "Mistuning Analyses of a Bladed Disk: Pole-Zero and Modal Approaches," GT2008 – 50191, ASME Turbo Expo, Berlin, Germany.

Sinha, A. 2008b, "Reduced-Order Model of a Mistuned Multi-Stage Bladed Rotor," *International Journal of Turbo and Jet Engines*, Vol. 25, pp. 145–153 (also, ASME GT 2007 – 27277).

Sinha, A., 2009, "Reduced-Order Model of a Bladed Rotor with Geometric Mistuning," *ASME Journal of Turbomachinery*, Vol. 131, 031007-1-7, July.

Sinha, A., 2010, *Vibration of Mechanical Systems*, New York: Cambridge University Press.

Sinha, A., and Bhartiya, Y., 2010, "Modeling Geometric Mistuning of a Bladed Rotor: Modified Modal Domain Analysis," *IUTAM Symposium on Emerging Trends in Rotor Dynamics,* K. Gupta (ed.), IUTAM Book Series, New York: Springer, pp. 177–184.

Sinha, A., and Chen, S., 1989, "A Higher-Order Technique to Compute the Statistics of Forced Response of a Mistuned Bladed Disk," *Journal of Sound and Vibration*, Vol. 130, pp. 207–221.

Sinha, A., Hall, B., Cassenti, B., and Hilbert, G., 2008, "Vibratory Parameters of Blades from Coordinate Measurement Machine Data," *ASME Journal of Turbomachinery*, Vol. 130, January, pp. 011013-1 to 011013-8 (also, ASME GT 2005 – 69004).

Song, S. H., Castanier, M. P., and Pierre, C., 2005, "Multi-Stage Modeling of Turbine Engine Rotor Vibration," DETC 2005 – 85740, Proceedings of IDETC/CIE 2005, ASME 2005 International Design Engineering Technical Conference, Long Beach, CA, September, pp. 1533–1543.

*Statistics Toolbox for use with MATLAB*, 2002, Natick, MA: The MathWorks, Inc.

Sternchuss, A., and Balmes, E., 2007, "Reduction of Multistage Rotor Using Cyclic Modeshapes," GT2007-27974, Proceedings of ASME Turbo Expo, Montreal, Canada.

Strang, G., 1988, *Linear Algebra and Its Applications*, 3rd ed., San Diego, CA: Harcourt Brace Jovanovich Publishers.

Tao, T., 2013, Private Communications, https://www.math.ucla.edu/~tao/.

Thompson, E. A., and Becus, G. A., 1993, "Optimization of Blade Arrangement in a Randomly Mistuned Cascade Using Simulated Annealing," AIAA 93 – 2254, Proceedings of AIAA/SAE/ASME/ASEE 29th Joint Propulsion Conference and Exhibit, Monterey, CA.

Vidyasagar, M., 1993, *Nonlinear Systems Analysis*, New Jersey: Prentice-Hall.

Vishwakarma, V., and Sinha, A., 2015, "Forced Response Statistics of a Bladed Rotor with Geometric Mistuning," *AIAA Journal*, Vol. 53, No. 9, pp. 2776–2781.

Vishwakarma, V., Sinha, A., Bhartiya, Y., and Brown, J.M., 2015, "Modified Modal Domain Analysis of a Bladed Rotor Using Coordinate Measurement Machine Data," *ASME Journal of Engineering for Gas Turbines and Power*, Vol. 137, No. 4, pp. 042502-1 to 042502-8.

Wagner, J. T., 1967, "Coupling of Turbomachine Blade Vibration through the Rotor," *ASME Journal of Engineering for Power*, Vol. 89, No. 4, October, pp. 502–512.

Wei, S.-T., and Pierre, C., 1988, "Localization Phenomena in Mistuned Assemblies with Cyclic Symmetry, Part I: Free Vibration," *ASME Journal of Vibration, Acoustics, Stress, and Reliability in Design*, Vol. 110, pp. 429–438.

Weiner, N., 1958, *Nonlinear Problems in Random Theory*, Cambridge, MA: MIT Press.

Weisstein, E. W., 2006, "Gram-Schmidt Orthonormalization," MathWorld – A Wolfram Web Resource, http://mathworld.wolfram.com/Gram-SchmidtOrthonormalization.Html.

Whitehead, D. S., 1960, "Force and Moment Coefficients for Vibrating Airfoils in Cascade," R. & M. No. 3254, *Aeronautical Research Council Report and Memoranda*, London, England.

Whitehead, D. S., 1966, "Effects of Mistuning on the Vibration of Turbomachine Blades Induced by Wakes," *Journal of Mechanical Engineering Science*, Vol. 8, No. 1, pp. 15–21.

Whitehead, D. S., 1998, "The Maximum Factor by Which Forced Vibration of Blades Can Increase Due to Mistuning," *ASME Journal of Engineering for Gas Turbine and Power*, Vol. 120, No. 1, pp. 115–119.

Xiangjun, Q., and Shijing, Z., 1986, "A Development of Eigenvalue Perturbation Method for Linear Discrete Systems," *Proceedings of the 4th International Modal Analysis Conference*, Los Angeles, CA, Vol. 1, pp. 556–562.

Xiao, B., Rivas-Guerra, A. J., and Mignolet, M. P., 2004, "Maximum Amplification of Blade Response Due to Mistuning in Multi-Degree-of-Freedom Blade Models," GT2004-54030, Proceedings of ASME Turbo Expo, Vienna, Austria.

Yang, M.-T., and Griffin, J. H., 1999, "A Reduced Order Model of Mistuning Using a Subset of Nominal System Modes," *ASME Journal of Engineering for Gas Turbines and Power*, Vol. 123, No. 4, pp. 893–900.

Yao, J., Wang, J., and Li, Q., 2009, "Robustness Analysis of Mistuned Bladed Disk Using the Upper Band of Structured Singular Value," *ASME Journal of Engineering for Gas Turbines and Power*, Vol. 131, pp. 032501-1 to 032501-7.

Zhang, Y.-Q., and Wang, W.-L., 1995, "Eigenvector Derivatives of Generalized Nondefective Eigenproblems with Repeated Eigenvalues," *ASME Journal of Engineering for Gas Turbines and Power*, Vol. 117, pp. 207–212.

# Index